Qualitative Data Collection Tools

Praise for this Book

"This is a very practical text for doctoral students learning research, in particular for those doctoral students in professional non-traditional doctoral degree programs who have never been exposed to the process of conducting qualitative research. The text brings students through the journey of qualitative data collection and provides a how-to approach to qualitative methodology."—**Emily Martinez Vogt, Florida Institute of Technology**

"Very useful and precise book on methods of research which could address all social scientific inquiry. It is a very well-researched research book and puts forth a thorough step-by-step process of doing publishable research."—**Abdy Javadzadeh, St. Thomas University**

"A comprehensive guide for the novice and even intermediate qualitative researcher. I highly recommend this easy-to-follow textbook for anyone considering a qualitative study."—**P. Bruce Uhrmacher, University of Denver**

"This new book fills a dire need in qualitative research—effectively and rigorously designing qualitative interview instruments and observation and artifact rubrics. The author has skillfully woven together necessary elements of qualitative design, research strategies, and research question development in order for researchers to operationalize their curiosities into successful, effective, and ethical research studies. I am particularly taken by the tone of the book. The author has written a book that challenges yet scaffolds readers to produce qualitative research studies with integrity and rigor."—**Karin Lindstrom Bremer, Minnesota State University**

"I would certainly adopt this book for my course. I think it offers a rich and much-needed perspective in the field of qualitative research."
—**Robert T. Palmer, Howard University**

"I think that the more exposure that students have to instrumentation the more informed that they can be when they build their own research studies."—**Tricia J. Stewart, Western Connecticut State University**

"I would love to purchase this text for myself. . . Additionally, I would highly recommend to students to purchase for both current and future use (templates for example could be used by these future practitioners in the clinic if conducting research)."—**Carol Lambdin-Pattavina, University of New England**

Qualitative Data Collection Tools
Design, Development, and Applications

Felice D. Billups
Johnson & Wales University

Los Angeles | London | New Delhi
Singapore | Washington DC | Melbourne

Qualitative Research Methods Series

Series Editor: David L. Morgan, *Portland State University*

The **Qualitative Research Methods Series** currently consists of 55 volumes that address essential aspects of using qualitative methods across social and behavioral sciences. These widely used books provide valuable resources for a broad range of scholars, researchers, teachers, students, and community-based researchers.

The series publishes volumes that

- Address topics of current interest to the field of qualitative research.
- Provide practical guidance and assistance with collecting and analyzing qualitative data.
- Highlight essential issues in qualitative research, including strategies to address those issues.
- Add new voices to the field of qualitative research.

A key characteristic of the Qualitative Research Methods Series is an emphasis on both a *"why to"* and *"how-to"* perspective, so that readers will understand the purposes and motivations behind a method, as well as the practical and technical aspects of using that method. These relatively short and inexpensive books rely on a cross-disciplinary approach, and they typically include examples from practice; tables, boxes, and figures; discussion questions; application activities; and further reading sources.

New and forthcoming volumes in the Series include:

Qualitative Longitudinal Methods: Researching Implementation and Change

Mary Lynne Derrington

Qualitative Data Collection Tools: Design, Development, and Applications

Felice D. Billups

Hybrid Ethnography: Online, Offline, and In-between

How to Write a Phenomenological Dissertation

Photovoice for Social Justice: Image Capturing in Action

Reflexive Narrative: Self-Inquiry Towards Self-Realization and Its Performance

For information on how to submit a proposal for the Series, please contact:

- David L. Morgan, Series Editor: morgand@pdx.edu
- Leah Fargotstein, Acquisitions Editor, SAGE: leah.fargotstein@sagepub.com

FOR INFORMATION:

SAGE Publications, Inc.
2455 Teller Road
Thousand Oaks, California 91320
E-mail: order@sagepub.com

SAGE Publications Ltd.
1 Oliver's Yard
55 City Road
London, EC1Y 1SP
United Kingdom

SAGE Publications India Pvt. Ltd.
B 1/I 1 Mohan Cooperative Industrial Area
Mathura Road, New Delhi 110 044
India

SAGE Publications Asia-Pacific Pte. Ltd.
18 Cross Street #10-10/11/12
China Square Central
Singapore 048423

Acquisitions Editor: Leah Fargotstein
Editorial Assistant: Claire Laminen
Production Editor: Vishwajeet Mehra
Copy Editor: Michelle Ponce
Typesetter: Hurix Digital
Proofreader: Jeff Bryant
Indexer: Maria Sosnowski
Cover Designer: Candice Harman
Marketing Manager: Shari Countryman

Copyright © 2021 by SAGE Publications, Inc.

All rights reserved. Except as permitted by U.S. copyright law, no part of this work may be reproduced or distributed in any form or by any means, or stored in a database or retrieval system, without permission in writing from the publisher.

All third party trademarks referenced or depicted herein are included solely for the purpose of illustration and are the property of their respective owners. Reference to these trademarks in no way indicates any relationship with, or endorsement by, the trademark owner.

Printed in the United States of America

ISBN: 978-1-5443-3482-0

This book is printed on acid-free paper.

20 21 22 23 24 10 9 8 7 6 5 4 3 2 1

Brief Contents

List of Tables	xv
List of Templates	xvi
Preface	xix
Acknowledgments	xxiii
Glossary of Terms	xxv
About the Author	xxvii

Chapter 1 The Qualitative Data Collection Cycle	1
Chapter 2 Using the Research Question to Guide Qualitative Data Collection Tool Design	15
Chapter 3 Conducting the Qualitative Study: Researcher Role, Access, Trustworthiness, and Ethical Concerns	23
Chapter 4 Interview Protocols	36
Chapter 5 Conversational and Discourse Analysis Tools	87
Chapter 6 Focus Group Moderator Guides	96
Chapter 7 Observation Tools	132
Chapter 8 Document and Artifact Analysis Tools	143
Chapter 9 Reflective Practice Tools	153
Chapter 10 Synthesis: The Qualitative Story	166

Appendices: A Case Study of Department X	175
Recommended Qualitative Research Websites	191
References	193
Index	205

Detailed Contents

A Guide To Qualitative Data Collection Tools

List of Tables	xv
List of Templates	xvi
Preface	xix
• Why Do We Need a Text on Qualitative Data Collection Tools?	xix
• Organization of the Text	xxi
Acknowledgments	xxiii
Glossary of Terms	xxv
About the Author	xxvii

Chapter 1 The Qualitative Data Collection Cycle — 1

What Is Qualitative Research?	1
Qualitative Research as a Worldview	2
Characteristics of Qualitative Research	3
Qualitative Research Designs	5
Qualitative Research Design Applications	9
The Sources of Qualitative Data	10
Qualitative Data Collection Tools	11
Highlights	13

Chapter 2 Using the Research Question to Guide Qualitative Data Collection Tool Design — 15

The Role of the Research Question in the Design of Qualitative Tools	15
Connecting Qualitative Designs With Guiding Questions, the Research Purpose, and the Research Questions	17
Connecting Qualitative Designs With Data Collection Strategies and Tools	20
Piloting Qualitative Tools	21
Highlights	22

Chapter 3 Conducting the Qualitative Study: Researcher Role, Access, Trustworthiness, and Ethical Concerns — 23

- The Researcher's Role in Qualitative Research — 23
- Researcher Bias and the Practice of Bracketing — 24
- The Researcher's Access to Data — 26
  - Research Sites — 26
  - Gatekeepers — 26
- Trustworthiness in the Qualitative Study — 27
- Ethics in Qualitative Research — 31
  - Ethical Concerns — 31
  - Institutional Review Boards and the Qualitative Study — 33
- Highlights — 35

Chapter 4 Interview Protocols — 36

- Interviewing Defined — 36
- Interview Applications — 37
- Requisite Skills and Characteristics of the Interviewer — 38
- Interview Formats and Types — 40
  - Formats — 40
  - Types — 42
- Getting Started With a Basic Template — 44
  - General Design Considerations — 44
  - Basic Interview Templates — 47
    - *Unstructured Interview Protocols* — 47
    - ▶ **Template 4.1:** Unstructured Interview Protocol — 48
    - *Semistructured Interview Protocol* — 50
    - ▶ **Template 4.2:** Semistructured Interview Protocol — 50
    - *Interviewer's Note-Taking Recording Sheet* — 52
    - ▶ **Template 4.3:** Interviewer Note-Taking Recording Sheet — 53
    - *Templates Variations: Interview Protocols for Specific Qualitative Designs* — 54
  - Phenomenological Interview Protocols — 55
    - *Phenomenological Interviewing Defined* — 55
    - *Phenomenological Interviewing Types and Approaches* — 55
    - *Defining Features of the Lived Experience and Essence Statements* — 56
    - *Phenomenological Interview Protocol* — 57

▶ **Template 4.4:** Phenomenological Lived Experience
  Interview Protocol   58
Ethnographic Interview Protocols   61
  Ethnographic Interviewing Defined   *61*
  Ethnographic Interviewing Types and Approaches   *61*
  Defining Features of the Ethnographic Interview:
  The Key Informant and the Researcher's Participation   *63*
  The Ethnographic Interview Protocol   *64*
  ▶ **Template 4.5:** Ethnographic Interview Protocol   65
Narrative Interview Protocols   67
  Narrative Interviewing Defined   *67*
  Narrative Interviewing Types and Approaches   *69*
  Defining Features of the Narrative Interview   *70*
  The Narrative Interview Protocols   *71*
  ▶ **Template 4.6:** Narrative Thematic/Structural/
    Dialogic Interview Protocol   72
  ▶ **Template 4.7:** Narrative Dramatism Interview Protocol   75
  ▶ **Template 4.8:** Narrative Life History/Life Story
    Interview Protocol   77

Piloting the Interview Protocol   80

Transforming Interview Data for Analysis   82

Highlights   85

Chapter 5 Conversational and Discourse Analysis Tools   87

Conversational and Discourse Analysis Defined   87

Conversational and Discourse Analysis Applications   89

Requisite Skills and Characteristics of the
Conversational and Discourse Analysis Researcher   89

Conversational and Discourse Analysis Formats   89

Getting Started With a Basic Template   90
  General Design Considerations   90
  Basic Conversational or Discourse Analysis Template   91
  ▶ **Template 5.1:** Conversational or Discourse Analysis Log   91

Template Variations and Challenges   93

Piloting Conversational and Discourse Analysis Tools   93

| Transforming Conversational and Discourse Data for Analysis | 94 |
| Highlights | 95 |

Chapter 6   Focus Group Moderator Guides — 96

Focus Groups Defined — 96

Focus Group Applications — 98

Requisite Skills and Characteristics of the Focus Group Moderator — 98

Focus Group Types and Variations — 99

Getting Started With a Basic Template — 101
   General Design Considerations — 101
      Developing Focus Group Moderator Guide Questions — *101*
      Sequencing Focus Group Moderator Guide Questions — *102*
      Developing Probes — *104*
      Using the Moderator's Guide to Conduct the Session — *105*
   Creating the Pre-Focus Group Profile Questionnaire — 105
   ▶ **Template 6.1:** Focus Group Moderator's Guide: Single Purpose — 106
   ▶ **Template 6.2:** Focus Group Presession Participant Profile Questionnaire — 108
   Focus Group Note-Taking Recording Sheet — 109
   ▶ **Template 6.3:** Focus Group Note-Taking Recording Sheet — 109

Template Variations: Focus Group Moderator Guides by Focus Group Type — 110
   ▶ **Template 6.4.1:** Focus Group Moderator's Guide for Two-Way Designs (Group #1) — 111
   ▶ **Template 6.4.2:** Focus Group Moderator's Guide for Two-Way Designs (Group #2) — 113
   ▶ **Template 6.5:** Focus Group Moderator's Guide for Dual Moderators — 116
   ▶ **Template 6.6:** Focus Group Moderator's Guide for Dueling Moderators — 118
   ▶ **Template 6.7:** Focus Group Moderator's Guide for Brainstorming/Envisioning — 120
   ▶ **Template 6.8:** Focus Group Moderator's Guide for Program Evaluation — 122
   Dyadic Interviews: The Facilitated 2-Participant Interview — 124
   ▶ **Template 6.9:** Moderator's Guide for Dyadic Interviews — 125

Piloting Focus Group Moderator Guides 127

Transforming Focus Group Data for Analysis 128
    Classic Approach 129
    Analytical Frameworks 130

Highlights 131

Chapter 7   Observation Tools   132

Observation Defined 132

Observation Applications 133

Requisite Skills and Characteristics of the Observer 134

Observation Formats 135

Getting Started With the Basic Template 136
    General Design Considerations 136

Basic Observation Templates 138
    ▶ **Template 7.1:** Observation Rubric for Formal or Informal Settings 138
    Template Variations and Challenges 139
    ▶ **Template 7.2:** Observation Log for Ethnographic Field Notes 139
    ▶ **Template 7.3:** Observation Rubric for Conceptual Frameworks 140

Piloting Observation Rubrics 141

Transforming Observation Data for Analysis 141

Highlights 142

Chapter 8   Document and Artifact Analysis Tools   143

Documents and Artifacts Defined 143

Document and Artifact Applications 144

Requisite Skills of Document and Artifact Recorders 145

Document and Artifact Types 146

Getting Started With a Basic Template 147
    General Design Considerations 147
    Basic Template 148
    ▶ **Template 8.1:** Combined Document/Artifact Rubric 148

Template Variations and Challenges	148
▶ **Template 8.2:** Artifact Rubric for Objects/Tangible Evidence	149
▶ **Template 8.3:** Artifact Rubric for Researcher Interpretations	150
▶ **Template 8.4:** Document/Artifact Rubric for Program Assessment/Evaluation	150
Piloting Document and Artifact Rubrics	151
Transforming Document and Artifact Data for Analysis	151
Highlights	152

Chapter 9 Reflective Practice Tools — 153

Reflective Practices Defined	153
Reflective Practice Applications	154
Requisite Skills of the Reflective Researcher and Participant Reflections	155
Reflective Format Types	156
Getting Started With the Basic Templates	158
General Design Considerations	158
Basic Reflective Questionnaire Templates	160
▶ **Template 9.1:** Single Topic Reflective Questionnaire	160
▶ **Template 9.2:** Free Association Reflective Questionnaire	161
▶ **Template 9.3:** Scenario Reflective Questionnaire	162
▶ **Template 9.4:** Reflective Journal and Diary Logs/Notebooks	162
Template Variations for Reflective Tools	163
Piloting Reflective Tools	163
Transforming Reflective Data for Analysis	164
Highlights	165

Chapter 10 Synthesis: The Qualitative Story — 166

A Multifaceted Enterprise	166
The Practitioner's Perspective	167
The Qualitative Design Process	168

A Plan for Action	170
Recommendations	172
Conclusion	174
Appendices: A Case Study of Department X	**175**
Appendix A Exemplar: Unstructured Interview Protocol	177
Appendix B Exemplar: Semistructured Interview Protocol	179
Appendix C Exemplar: Focus Group Moderator's Guide: Single Purpose	181
Appendix D Exemplar: Observation Rubric for Formal or Informal Settings	184
Appendix E Exemplar: Combined Document/Artifact Rubric	186
Appendix F Exemplar: Single Question Reflective Questionnaire	188
Appendix G Exemplar: Data Collection Plan	189
Recommended Qualitative Research Websites	**191**
References	**193**
Index	**205**

List of Tables

1.1 Qualitative Research Designs, Guiding Questions, & Design Characteristics	7
1.2 Sources of Qualitative Data	10
1.3 Qualitative Data Collection Tools	12
2.1 Qualitative Research Designs, Guiding Questions, & Research Purpose Statements	17
2.2 Qualitative Research Designs, Research Questions, & Keywords	19
2.3 Qualitative Research Designs, Data Collection Strategies, & Tools	20
3.1 Commonalities Between Quantitative and Qualitative Research for Rigor	28
3.2 What Do Institutional Review Boards Want to Know About Your Study?	33
4.1 Qualitative Designs/Approaches Using Semi-Structured or Unstructured Interview Protocols	54
4.2 Qualitative Designs Where Customized Interview Protocols Are Recommended	54
6.1 Focus Group Types	100
6.2 Moderator Guides by Focus Group Variation/Type	110
8.1 Document and Artifact Types	144
9.1 Reflective Practice Tools, Purposes, & Qualitative Designs	157
10.1 The Qualitative Design Process	169
10.2 The Qualitative Data Collection Plan (SAMPLE)	171

List of Templates

4.1 Unstructured Interview Protocol	48
4.2 Semistructured Interview Protocol	50
4.3 Interviewer Note-Taking Recording Sheet	53
4.4 Phenomenological Lived Experience Interview Protocol	58
4.5 Ethnographic Interview Protocol	65
4.6 Narrative Thematic/Structural/Dialogic Interview Protocol	72
4.7 Narrative Dramatism Interview Protocol	75
4.8 Narrative Life History/Life Story Interview Protocol	77
5.1 Conversational or Discourse Analysis Log	91
6.1 Focus Group Moderator's Guide: Single Purpose	106
6.2 Focus Group Presession Participant Profile Questionnaire	108
6.3 Focus Group Note-Taking Recording Sheet	109
6.4.1 Focus Group Moderator's Guide for Two-Way Designs (Group #1)	111
6.4.2 Focus Group Moderator's Guide for Two-Way Designs (Group #2)	113
6.5 Focus Group Moderator's Guide for Dual Moderators	116
6.6 Focus Group Moderator's Guide for Dueling Moderators	118
6.7 Focus Group Moderator's Guide for Brainstorming/Envisioning	120
6.8 Focus Group Moderator's Guide for Program Evaluation	122

6.9 Moderator's Guide for Dyadic Interviews — 125

7.1 Observation Rubric for Formal or Informal Settings — 138

7.2 Observation Log for Ethnographic Field Notes — 139

7.3 Observation Rubric for Conceptual Frameworks — 140

8.1 Combined Document/Artifact Rubric — 148

8.2 Artifact Rubric for Objects/Tangible Evidence — 149

8.3 Artifact Rubric for Researcher Interpretations — 150

8.4 Document/Artifact Rubric for Program Assessment/Evaluation — 150

9.1 Single Topic Reflective Questionnaire — 160

9.2 Free Association Reflective Questionnaire — 161

9.3 Scenario Reflective Questionnaire — 162

9.4 Reflective Journal and Diary Logs/Notebooks — 162

This book is dedicated to my husband, Tom, and my daughter, Moriah, who always support me with love.

Preface

Numerous texts on qualitative research, covering all aspects of designing, collecting, and analyzing qualitative data, are available to researchers, faculty, and students. These texts introduce readers to the nature and characteristics of qualitative research, qualitative research designs, specific strategies for designing qualitative studies, and qualitative data analysis methods. The one resource, however, that is missing from this list is a compendium on how to design, develop, test, and employ qualitative tools in order to collect qualitative data. This omission leaves qualitative researchers at a loss as they attempt to create effective qualitative tools for their projects.

Why Do We Need a Text on Qualitative Data Collection Tools?

This niche in the qualitative research market has become increasingly apparent to me over the past 30 years. As a researcher and a teacher, I have continually searched for a comprehensive resource on the design of qualitative tools. I believe that this gap in the market compromises the efforts of qualitative researchers, who must resort to creating their own tools with minimal guidance or examples. In fact, what typically happens is that qualitative researchers adapt existing tools from other projects that may not adequately suit the research approach they are employing.

When students are assigned a project to develop protocols, guides, logs, and/or rubrics, they often succumb to an Internet search in order to adapt someone else's design. This action often results in a misguided appropriation of the tool with the intended research design. In many cases, searching online for information about qualitative data collection tools merely leads the researcher to topics such as how to develop survey questionnaires based on qualitative data results, how to develop interview or focus group guides (to the exclusion of any other type of instrument), or how to analyze data that have already been collected. In all of these cases, the information misses the mark.

The development of effective instrumentation must be grounded in more than merely replicating examples located on the Internet. A good design must originate from a clear research objective and a reasonable research design, applicable to the problem under study. For instance, students often mistakenly think that simply making a list of questions will suffice when creating an interview protocol; they often consider that 20 or 30 questions are manageable for a focus group, without consideration for the group's synergy or the timing of the discussion. Similarly, students often fail to understand that for every data source, a well-designed tool is necessary to collect and organize the data. Data collection plans that include observation, field notes, journaling, document and artifact analysis, conversational and discourse analysis—in addition to the more commonly employed interviews or focus groups—require specific applications. In spite of their prevalence in qualitative studies, examples of data collection tools are nearly impossible to locate.

Therefore, as the student struggles, so does the instructor. A resource that supports both the teacher and the student learner would remove many of the current challenges for the research methods courses offered in so many programs. My experience, directly and in consultation with colleagues, is that we all work around the existing textbooks on qualitative research and extract what we can from those texts to patch together the information we need to instruct ourselves and our students. A resource that provides templates and instructions for how to develop these tools would be a welcome addition to the resources currently on the market.

It is important to note that this text is *not* intended to do certain things. It will not provide a detailed explanation of the history, properties, or worldview perspective of qualitative research; these topics are comprehensively covered in other excellent texts. Similarly, this text will not discuss the specifics of each qualitative research design other than a reference to how that research design affects the design and development of the requisite tools. Data analysis and data management will not be the focus of this volume, although a description of how data are collected by using the appropriate tool is a prelude to preparing data for analysis. It is not meant to be read from cover to cover but rather used as a resource for designing specific tools to match specific research applications. It is meant to serve as a starting point for researchers who want to use these templates and customize them to match their research objectives. This volume will, therefore, focus solely on the basis for and guidance related to the design and development of qualitative data collection tools.

Thus, four goals ground this project:

1. To create a practical guide and compendium that will fill a gap in the qualitative research methods literature regarding the

conceptualization, design, development, testing, and application of qualitative tools for specific qualitative research strategies

2. To provide templates and exemplars, from which readers may modify their own tools

3. To create a companion piece to support existing research methods texts and to provide resources to help them customize their tools to match their research designs

4. To provide a resource for researchers, faculty, and students who require easy-to-understand information in support of learning, teaching, and/or conducting qualitative research in education, the social and behavioral sciences, marketing/business, and in the health care fields

Organization of the Text

The text opens with a glossary of terms in order to establish a common understanding of the labels used for the qualitative tools discussed herein; some of these labels are more established in the qualitative lexicon than others, and some of the labels may even be considered interchangeable. The Glossary provides a starting point for the discussion in the subsequent chapters.

Next, the chapters are divided into two main sections:

- Chapters 1, 2, and 3: Context for Conducting Qualitative Studies. Chapter 1 provides the reader with an overview of the characteristics of qualitative research, the most commonly used qualitative research designs and their applications in a study, and the sources of and tools for collecting qualitative data. Chapter 2 reviews the role of the research purpose and research questions to guide the design of qualitative data collection tools, connecting these designs with the appropriate data collection strategies. Chapter 3 covers the researcher's role in the qualitative research project, including the concepts of bias, the practice of bracketing, and the researcher's access to research sites. Trustworthiness strategies and ethical concerns for conducting qualitative studies close the chapter.

- Chapters 4, 5, 6, 7, 8, and 9: Designing, Developing, and Piloting Qualitative Tools. These chapters constitute the primary purpose of the text, and each chapter focuses on a different qualitative toolkit, including any variations in design and application. First, each qualitative strategy for which the tool is designated is defined for the reader, followed by the typical applications of the tool's use, the requisite skills required of the researcher who intends to use the tool, their various formats and/or types, and the design considerations for how to develop the basic tool. In many cases, there are also variations for each of these basic formats, and examples are provided. Each of the instrumentation chapters concludes with a brief description of how to pilot (pre-test) the tools, and how to prepare the resulting data for analysis.

More importantly, every instrumentation chapter includes detailed templates to match that chapter's data collection focus. These templates form the main purpose of the text, and they provide the reader with a starting point from which to customize and refine their own cadre of tools to match their research projects. The templates have been designed with this refinement in mind, ensuring that every researcher will be able to adapt the templates to help them collect the data they need. Exemplars for select templates have been provided in the Appendix to provide a framework for researchers as they customize their own tools.

Chapter 10 highlights the qualitative research project as an integrative process. The right research design, matched with the right data sources and the right instrumentation, will yield the best possible results. It is in those results that the power of the qualitative story emerges. This chapter also reviews how different categories of researchers might use qualitative tools for their projects (student, faculty, researcher), revisits the value of conducting the pilot test and evaluating the tools for a study, and offers a plan for organizing a study and the tools used in a qualitative project.

It is my hope that all researchers, instructors, and students, whether novice or expert, will find this volume a useful and credible supplement to other qualitative research resources. Studying the experiences, interactions, and worlds of individuals from their own perspectives is the essence of the qualitative exploration, but without the right tools, those phenomena remain elusive and irretrievable. This guide is just one small step toward adequately equipping the qualitative researcher.

Acknowledgments

I am grateful to the many individuals who supported me in the production of this text. First, and foremost, I am indebted to my doctoral students, who repeatedly lamented the absence of qualitative research instrument examples in the literature. Slowly, and together, we developed the rudimentary beginnings of qualitative data collection instrumentation. This volume exists because of their work, their dedication, and their pursuit of the necessary tools for their own research.

Second, this work reflects the pioneering qualitative researchers and scholars who have promoted this research approach over the years. As a new doctoral student in the late 1980s who pursued a qualitative dissertation project, I found that qualitative research was viewed with suspicion by many faculty. Since that time, qualitative research has evolved and taken hold in the social sciences with greater legitimacy and dominance. Many of the works referenced in this text are the works that grounded this movement from those early beginnings. I hope I have acknowledged their collective contributions with all the justice they deserve.

Third, this work would not have been possible without the incredible assistance of several individuals. Dr. Robert K. Gable, EdD, has served as my primary supporter, cheerleader, and advisor from the moment I first conceived of this project nearly six years ago. Dr. Jennifer Broderick, EdD, has served as my partner in preparing and reviewing the manuscript and has been a faithful colleague and friend throughout the process; her sharp eye never missed a thing as she reviewed my drafts. Dr. Christine Perakslis, EdD, my guiding light and kindred spirit, never once doubted my work, my vision for this project, or in my ability to complete it. Her belief in me kept me going through many periods of doubt and delay.

Finally, my profound thanks goes to Leah Fargotstein, Acquisitions Editor at SAGE, who remained calm, composed, and encouraging through all the phases of this project. New to book writing, I required considerable assistance; she was supportive from the first iteration of the project to the last and has taught me a great deal about the ebb and flow of the publication process. I have benefitted tremendously from her counsel and coaching. Along with Leah, Dr. David Morgan provided the guidance and valuable feedback that helped me refine this text; it is an honor for me to benefit from one of the experts whom I have referenced over the years. Additionally, thank you to the reviewers who carefully scrutinized my manuscript drafts and helped me refine and produce a final document that is far better than anything I could have produced on my own. If it is true that it takes a team to complete a project of this sort, I have benefited from the guidance of a wise editors, insightful reviewers, and committed friends and colleagues. I am indebted to you all.

Felice D. Billups, June 2019

The author and SAGE would like to thank the following reviewers for their feedback during development:

- Karen J. Aroian, University of Central Florida
- Andrea M. Flynn, Concordia University Chicago
- Abdy Javadzadeh, St. Thomas University
- Carol Lambdin-Pattavina, University of New England
- Karin Lindstrom Bremer, Minnesota State University, Mankato
- Jacqueline Lynch, Florida International University
- Gabriela Novotna, University of Regina
- Robert T. Palmer, Howard University
- Juliana Svistova, Kutztown University
- Tricia J. Stewart, Western Connecticut State University
- P. Bruce Uhrmacher, University of Denver
- Emily Martinez Vogt, Florida Institute of Technology
- Robyn Williams, Charles Darwin University
- Julie Zadinsky, Augusta University
- Lauren Zucker, Fordham University

Glossary of Terms

Diary

A record of events, transactions, or observations documented daily or at frequent intervals; a record of personal activities, reflections, and feelings (Chapter 9).

Field Notes

Written or recorded observations, thoughts, and impressions of the researcher's views on a research setting and its participants (Chapter 7).

Journal

A record of experiences, ideas, and reflections documented regularly for oneself or for another individual/organization; an account of day-to-day events maintained systematically, often focused on reflexive practices or research settings (Chapter 9).

Log/Logbook

A record of a performance, event, day-to-day activity, or interactions between individuals/groups at a site; to make an official record of something (Chapter 5).

Moderator's Guide

A device that organizes focus group topics and the questioning routes; to guide the group discussion and encourage group synergy to activate and generate new threads on the topics that are initially introduced (Chapter 6).

Notebook

A volume for recording notes or memos; a book with blank pages for recording notes, ideas, observations, or drawings (Chapter 9).

Protocol

A form or schedule that includes a set of questions (typically for interviews) related to the specific aims of a study; these questions guide the interviewer and direct the conversation for depth interviews (Chapter 4).

Questionnaire

An organized set of questions devised to provide data to support a study; in qualitative research projects, these questionnaires consist of open-ended questions and prompts (Chapter 9).

Recording Sheets

Sheets that capture recordings, writing, and note-taking related to conversations and observations; a vehicle to capture evidence and documentation about something (Chapter 4, 6).

Rubric

A guide or chart listing specific criteria and/or categories for assessing, evaluating, scoring, grading, or categorizing something (Chapter 7, 8).

Tools

Devices used for careful and exact work in order to accomplish something of value; implements, devices, or instruments used to assist in the creation or production of something else (all chapters).

Sources

Schwandt, T. (2015). *The Sage dictionary of qualitative inquiry* (4th ed.). Thousand Oaks, CA: Sage.

Webster's Dictionary: https://www.merriam-webster.com/dictionary/dictionary

About the Author

Felice D. Billups is a professor in the Educational Leadership doctoral program at Johnson & Wales University in Providence, Rhode Island. She teaches courses in educational research and organizational behavior, specializing in qualitative and mixed methods research applications, and organizational culture studies. As a former college administrator, she has directed programs in strategic planning, institutional research and effectiveness, regional and specialized accreditation processes, and academic program reviews.

CHAPTER 1

The Qualitative Data Collection Cycle

> *Qualitative inquiry, by nature, is a customized, inductive, emergent process. . . . It means purposely adopting different lenses, filters, and angles as we view social life so as to discover new perceptions and cognitions about the facet of the world we're researching.*
>
> (Saldana, 2015, pp. 3–4)

What Is Qualitative Research?

For decades, scholars and researchers have struggled to define qualitative research. Patton (2015), Merriam (2002), and Maxwell (2005, 2013) agree that in its most fundamental form, qualitative research explores peoples' lives, behaviors, emotions, and perceptions. This definition does not, however, take us far enough into the intricacies of qualitative inquiry. From the qualitative, interpretive lens of viewing phenomenon, the focus of attention for qualitative research must revolve around the individual and unique experiences of the participants. As many scholars note, the key question in any qualitative exploration is this: What is really going on here?

It may be useful to consider a variety of definitions as the best way to understand the scope of the qualitative research approach. Denzin and Lincoln's seminal definition is presented as a starting point:

> Qualitative research is a situated activity that locates the observer in the world. Qualitative research consists of a set of interpretive, material practices that make the world visible. These practices transform the world. They turn the world into a series of representations, including field notes, interviews, conversations, photographs, recordings, and memos to the self. At this level, qualitative research involves an interpretive, naturalistic approach to the world. (2011, p. 3)

Researchers support this definition by adding their own viewpoint, including Patton's (2015) definition that captures the intimacy of interaction between researcher and participant, and Creswell and Poth's (2018) and Saldana's (2015) definitions, which speak to the power of multiple perspectives. The sum total of these definitions implies that qualitative research allows us to uncover the meaning individuals ascribe to their experiences, through close interactions, rich conversations, and multifaceted interpretations.

Qualitative Research as a Worldview

What do we mean by a worldview when we talk about the different research approaches? Guba's definition (1990) is still the best and most succinct one, noting that a worldview is "a basic set of beliefs that guide action" (p.17). Another way to interpret a worldview is to understand it as a philosophy, a belief about how the world is ordered, or how reality or truth is perceived.

A worldview, then, is described through the lens of five basic assumptions: ontological, epistemological, axiological, rhetorical, and methodological. Each assumption refers to a different aspect of a research approach and references that approach to reality, the researcher's role, values and bias, the use of language, and the orientation for conducting research. In the qualitative perspective, as distinguished from the quantitative or mixed methods approaches, these assumptions are very different. In recent years, qualitative inquiry has been labeled as social constructivist, implying an approach to research that supports the multiple views and perspectives elicited from participants. This label compares with the positivist label for quantitative research that suggests a traditional, empirical approach to research where there is a single truth or reality. On the midpoint of this methodological continuum lies the mixed methods pragmatic approach, which combines the strengths of quantitative and qualitative methods into a single research study (Guba, 1990; Lincoln & Guba, 1985).

With regard to qualitative research, it is important to clarify the worldview assumptions in order to understand the nature of the inquiry. First, as the ontological perspective refers to the researcher's view of reality, the qualitative researcher positions reality as subjective, incorporating the multiple realties represented by participants. Second, the epistemological assumption refers to the researcher's role, which is intimate and interactive in qualitative studies (otherwise labeled as the "researcher as the data collection instrument"; Denzin & Lincoln, 2003). Third, the axiological assumption refers to the values in the qualitative approach, which are inherently biased and subjective, focused on the particularity of the case (Stake, 1995). Fourth, the rhetorical assumption refers to the use of language in qualitative inquiry, which infers that language is often framed in the first person, as a story or direct experience, and is informal, descriptive, and personal. Fifth, and finally, the methodological assumption refers to the naturalistic process for conducting research, which is inductive, holistic, and depends on triangulation of multiple data sources to corroborate findings. This overview of the worldview assumptions leads to a summary of the characteristics of qualitative research (Bernard, 2013; Bogdan & Bilken, 2003; Czarniawska, 1997; Stebbins, 2001).

Characteristics of Qualitative Research

Qualitative research is distinguished from the quantitative or mixed methods approaches by a grounding in the social constructionist worldview described previously. Scholars identify a set of characteristics that reflect the qualitative approach, as listed below (Creswell & Poth, 2018; Maxwell, 2013; Patton, 2015).

Natural setting. Research is conducted in a natural setting, a setting indigenous to the participants, rather than in a controlled or contrived setting that may be designed to reduce bias or extraneous factors; face-to-face interactions allow participants to provide their perspectives in the same setting where they experience the phenomenon and where it is familiar enough to offset any feelings of isolation or conflict.

Purposeful sampling. Participants are selected intentionally, chosen for their capacity to provide detailed information, based on their unique experiences and perspectives. Qualitative participants are often known as "information-rich" cases (Patton, 2015, p. 53).

Multiple data sources. A variety of data sets, accessed from different participant perspectives and experiences, are intentionally collected and corroborated to provide a holistic picture of an experience or phenomenon. Socially constructed reality, realities derived from the individuals selected for the study, provide the many viewpoints representative of an experience (Weller & Romney, 1988). In this way, verification and triangulation allows for the holistic picture of the phenomenon to emerge.

Interpretive experiences. The nature of qualitative data is interpretive, qualified, and expressive, captured in the words, stories, images, artifacts, and behaviors of the participants. Meaning is assigned to every word, story, behavior, and symbol in order to develop a comprehensive profile of the phenomenon.

Unique perspectives. The participants' meanings and interpretations are paramount, and their unique perspectives are represented in such a way as to protect the integrity of their views while acknowledging the varied viewpoints of the participants who share in the same experiences or phenomenon.

Holistic. Qualitative studies are interpretive and holistic, reflecting and extending the complex picture of a particular problem or issue, and delving deeply into the views and voices of the participants.

Emergent design. The qualitative design evolves over the course of the study, a design that is grounded in the researcher's original intuition, prior research studies, and an educated assessment of the phenomenon to be explored. This design process guides the project's development and should be refined and solidified as the study evolves.

Frameworks. A theoretical lens or framework often guides the qualitative project. Theory can be applied to a study in order to develop the research purpose, research questions, instrumentation, or to frame the research findings. Conceptual frameworks are often developed to organize and explain how theory is operationalized for the qualitative study. The role of theory in a qualitative design differs significantly from its role in a quantitative project, since it is not applied deductively in order to prove or test the theory; however, in the case of grounded theory designs, researchers may use extant theory as a starting point to develop a new working theory grounded from qualitative data or to explore specific elements of a theory from a qualitative perspective. Alternately, in a phenomenological study, the elements of several different theories may be operationalized in a conceptual framework to guide the design and implementation of data collection tools and an interpretation of the findings.

Researcher as instrument. The researcher is the primary conduit to data collection, otherwise stated as "researcher as key instrument" (Denzin & Lincoln, 2003), meaning that the distance between the qualitative researcher and the participants in a qualitative study is close, interactive, and openly subjective.

Inductive exploration. Finally, the nature of the qualitative research process is inductive, meaning that the study works from "the data of specific cases to a more general conclusion" (Schwandt, 2015, p. 153).

Given these qualitative worldview assumptions and characteristics, a researcher must effectively capture the complex, processual, rich views of participant stories and experiences. Therefore, a researcher must identify the data collection strategies that will uncover these stories and experiences sufficiently and clearly. The bridge that connects the participant voice and the data is the qualitative tool.

Qualitative Research Designs

While there are many variations in the types and labels for qualitative research designs, most scholars would agree on a common cadre of basic designs (Crabtree & Miller, 2015; Creswell & Poth, 2018; Denzin & Lincoln, 2013; Maxwell, 2013; Merriam, 2002; Patton, 2015; Silverman, 2013).

Descriptive/Interpretive. Descriptive/interpretive (the terms are used interchangeably for the purpose of this discussion) designs focus on how participants make meaning of a situation or phenomenon, where the researcher describes the collective experiences and seeks to discover or understand the participants' points of view (Merriam, 2002). This design attempts to answer the question of "What is . . . ?" rather than seeking to uncover a lived experience, an in-depth assessment of a process or event, or the narrative story of an individual or individuals. These designs are guided by the question, "How can we understand a participant's experience through his or her self-constructed meaning of the phenomenon under study?"

Phenomenological. While all qualitative research focuses on phenomenon, phenomenological designs explicitly focus on the essence of the lived experience, grounded in a shared human condition. For instance, the experience of this shared phenomenon may represent the human experiences of trauma, grief, joy, birth, death, illness, or healing. Attempts to deal with the individuals' inner experiences as they live through these phenomena help the researcher uncover the unexplored or subconscious aspects of those experiences. As a result, an "essence meaning" is created by synthesizing the collective lived experiences of participants in an attempt to represent their emotional, psychological, and transformative journeys (Colaizzi, 1978; Giorgi, 1985; Moustakas, 1994; van Manen, 2014). These designs are guided by the question, "What is the essence of the lived experience under study?"

Ethnographic. With roots in anthropology, ethnographic designs are defined as the substantive, analytical description of an intact cultural group in its natural setting. The researcher conducts field work to observe, record, interact, and dissect the various levels of the cultural activity, with the goal of understanding the how and the why of a cultural group's purpose and functioning. Akin to cultural analysis, the researcher uses continuous observation and reflection to record virtually everything that occurs in the research field. Participant observation is the most common method, where the researcher can obtain the insider's point of view (Fetterman, 2010; Hammersley & Atkinson, 2007; Schwartzman, 1992; Stewart, 1998; Van Maanen, 2011). These designs are guided by the question, "How can we study, uncover, and understand the intact culture of this group?"

Narrative. Narrative designs comprise the synthesis of individual stories reflecting an event or series of events, chronologically connected by the researcher. The focus is on the study of one or two individuals, and the meaning of their stories is embedded within the context of a larger phenomenon or cultural context. These stories (often called life histories or life stories) are further validated as an exploration of the social, cultural, familial, linguistic, and institutional narratives within which the individual experiences were constructed. Narratives focus on a unique story as the object of inquiry in order to determine how people make sense of the events in their lives; the researcher's challenge is to create a chronological record of the events from the narrative perspectives and to represent that story as a synthesized product (Atkinson, 2016; Clandinin, 2013; Gubrium & Holstein, 2003; Riessman, 2008). As Riessman (1993) notes, "narratives are essential meaning-making structures" (p. 4). Narrative explorations are often included as subsets of other qualitative designs, such as historical or ethnographic designs (Creswell & Poth, 2018). These designs are guided by the question, "What does this story reveal about this individual(s) and his or her (their) world(s)?"

Case Study. Case study designs are essentially situational analyses where a particular event, process, or setting is studied from the viewpoints of all key stakeholders. Through this situational analysis, the viewpoints of all stakeholders are integrated; the findings provide an intricate, collective perception that contributes to understanding the phenomenon under study. This deep exploration, where multiple sources of data are collected and corroborated, leads to a comprehensive understanding of how an event, process, or setting emerged, unfolded, succeeded, failed, or impacted a group or organization. Studying cases from multiple perspectives lends a richness and a multidimensional picture of how people function within organizational or historical incidents. Although some scholars position case studies to include the bounded study of individuals in a particular circumstance (Creswell & Poth, 2018; Patton, 2015; Thomas, 2015), many qualitative scholars refer to case study designs as the study of a process, event, setting, or circumstance bounded by time and context (Hamel, Dufour, & Fortin, 1993; Merriam & Tisdell, 2015; Stake, 1995). Therefore, these designs are essentially guided by the question, "How do stakeholders describe this process or event, and what does it tell us about future practice(s)?"

Grounded Theory. Grounded theory designs move beyond description to generate or discover an emergent theory, captured in a schema or visual diagram that displays the process that participants have experienced (Strauss & Corbin, 1998). The theory would explain the process, practice, or personal transition that provides the researcher with a framework for further research. The working or emergent theory is grounded in the data that originates from participants who have experienced the common practice, process, or transition (Birks & Mills, 2011; Charmaz, 2014; Clarke, Friese, & Washburn, 2017; Corbin & Strauss, 2015; Glaser, 2000; Glaser & Strauss, 1967; Strauss & Corbin, 1998). These designs are guided by the question, "What theory emerges from the systematic, comparative analysis of data originating from participants sharing the same experience?"

Historical. Historical designs are not always viewed as a form of qualitative research, but they embody all the characteristics and strengths of the qualitative approach. The historical approach is an analytical one, with various subdesigns (Brundage, 2017; Gall, Gall, & Borg, 2006; Lange, 2012; McDowell, 2013) that focus on specific elements of the

Table 1.1 Qualitative Research Designs, Guiding Questions, & Design Characteristics

Design	Discipline roots	Guiding Question	Characteristics
Descriptive/ Interpretive	Social Sciences, Humanities, Sociology	How can we understand a participant's experience through his/her self-constructed meaning of the phenomenon under study?	Exploring phenomenon from the participant's perspective

(Continued)

Table 1.1 (Continued)

Design	Discipline roots	Guiding Question	Characteristics
Phenomenological	Psychology, Social Psychology, Philosophy	What is the essence of the lived experience under study?	Exploring the lived experience, the essence of combined perspectives
Ethnographic	Anthropology, Sociology	How can we study, uncover, and understand the intact culture of this group?	Field studies, cultural exploration, and analysis to uncover the layers of meaning and activity within an intact cultural group
Narrative	Psychology, Literature	What does this story(ies) reveal about this individual(s) and their world(s)?	Revelations about key individuals and their personal stories
Case Study	Psychology Law, Political Science, Health Sciences	How do stakeholders describe this process/event/setting; what does it tell us about future practice(s)?	Participant-constructed meaning around a bounded event, process, or setting using multiple data sources
Grounded Theory	Psychology, Sociology	What theory emerges from the systematic, comparative analysis of data originating from participants sharing the same experience?	Developing working theory grounded in the data where systematic analysis is generated

Design	Discipline roots	Guiding Question	Characteristics
Historical	History, Ethics	How does the analysis of past events or lives of pivotal individuals inform us about the present or future state of things?	The analysis of past events to understand the present or project what might be best for the future

guiding question, "How does the analysis of past events or lives of pivotal individuals inform us about the present or future state of things?"

Qualitative Research Design Applications

When should you use a qualitative research design? According to scholars, there are several instances when the inductive, interpretive approach inherent in a qualitative design is ideal (Crabtree & Miller, 2015; Creswell & Poth, 2018; Denzin & Lincoln, 2013; Flick, 2009; Maxwell, 2013; Patton, 2015; Silverman, 2008):

- To explore an idea or topic
- To explore a process or event or phenomenon
- To gain insight into a group's culture, lifestyle, and history, as well as their motivations, behaviors, and preferences
- To further understand processes or events from multiple perspectives
- To supplement quantitative research findings or support the design of a mixed methods project

Conversely, there are several instances when a qualitative design is *not* recommended:

- To measure, investigate, or examine relationships, differences, comparisons, or causes
- To identify causal relationships
- To conduct an experiment
- To test a theory or a hypothesis(es)

In a research environment that remains predominantly quantitative, and where measurable attributes, causation, and quantifiable findings are acknowledged as reliable facts, qualitative research offers an alternative perspective. The strength of qualitative exploration lies in the holistic, interpretive uncovering of the human experience, reinforced by the stories and meanings individuals give to those experiences. There is much to be gained from this approach, as it leads us to explore how qualitative researchers obtain and integrate qualitative data for research projects.

The Sources of Qualitative Data

Where does qualitative data come from? Many new qualitative researchers may assume that participants' words are the only source for qualitative data and that interviews and focus groups are the only way to capture those words. In fact, this narrow view prevails among novice as well as seasoned researchers who have minimal experience with qualitative methods. The notion that the sources of qualitative data are rich and varied, and that the strategies for collecting qualitative data include numerous tools and techniques, often surprises individuals who are designing their own qualitative projects.

This overview addresses these perceptions by diagramming the rich, varied sources of qualitative data and the various tools that researchers might use to maximize this research approach. Beginning with a list and description of the qualitative data sources and followed by a list of the tools that researchers can use to capture these data, the qualitative data collection process becomes more transparent. The sources of all qualitative data are derived from the following activities, as outlined in Table 1.2.

Table 1.2 Sources of Qualitative Data	
Words, Conversations, Stories	People orally describe their experiences or perspectives, either alone or in groups
Conversational/Discourse Interactions	Researcher-as-participant/nonparticipant observes and analyzes the meaning of ongoing conversations where people communicate during their social interactions

Synergistic Discussions	Facilitated, structured group discussions where the researcher guides the group toward coordinated engagement as participants share their perspectives, opinions, experiences
Dyadic Interactions	Researcher-facilitated two-person synergistic conversations
Observations	Researcher observes and records the nonverbal and contextual behaviors and interactions of individuals or groups in formal or informal settings
Documents and Artifacts	Researcher reviews, records, and analyzes the meaning of contextual, extant documents, artifacts, cultural materials, and other tactile objects, often to support other data sources; in some cases, the participants generate the documents and artifacts such as in photo voice strategies or through windshield or walking interview recordings
Journals, Diaries, Reflections	Reflective devices constructed by researcher or participants to reflect on the focus of inquiry, to supplement primary data sources, and to debrief from the experience of sharing experiences with a researcher/peers

Qualitative Data Collection Tools

Qualitative data collection is labor-intensive, focused, and complex. As a qualitative researcher, you must plan to immerse yourself in the field for sufficient time to collect extensive data, understand the context for that data, and uncover the nuances of what is occurring. The concept of "seeing versus looking" is an essential skill that qualitative researchers must develop. In addition to what the senses can capture (hearing, seeing, feeling), a researcher must cultivate their sense of intuition and judgment. For instance, what do the nonverbal and contextual clues offer in the way of deeper understanding of the phenomenon?

Given these parameters, it is important to identify the different qualitative tools and the types of data they collect, as outlined in Table 1.3.

Table 1.3 Qualitative Data Collection Tools

Interview Protocols	A range of protocol types linked with specific approaches and research designs, used to guide a conversation
Conversation/Discourse Logs	Tools to capture the dynamics of conversation and interactions among participants in order to interpret the meaning of those conversations
Focus Group Moderator Guides	Constructed guides that direct the synergistic discussions of a group, allowing for sufficient structure to guide the conversation but leaving room for the group to direct the sequence of topics
Observation Rubrics	Tools that help the researcher record information from several different perspectives: what is observed, what is heard in dialogue, reflective notes in a journal from the researcher's point of view, and demographic profile notes about the time, place, and date of the field setting
Document and Artifact Rubrics	Researcher or participant constructed documents that allow for the categorization of documents and artifacts in order to compare, corroborate, and analyze that data in the context of a study

Reflective Tools	Journals, diaries, and reflective questionnaires as tools to collect reflective and reflexive data in a study where the data are either primary or secondary and either generated by the participant or the researcher
Supplemental Tools	Interviewer or focus group recorder sheets, prefocus group profile questionnaires, and other types of data collection tools that support primary data collection in a qualitative study

HIGHLIGHTS

The Qualitative Data Collection Cycle	Qualitative research defined: Exploration of people's lives, lived experiences, behaviors, emotions, experiences, feelings, perceptions, and interactions
	Qualitative worldview: Subjective reality based on multiple perspectives of purposefully selected participants, situated in their natural settings, framed holistically
	Characteristics of qualitative research: Natural setting, researcher as instrument, multiple data sources, rich and deep data collection, interpretive and socially constructed, emergent design, inductive inquiry
	Basic qualitative research designs: Descriptive/interpretive, phenomenological, ethnographic, narrative, case study, grounded theory, historical
	Design applications: Exploring an idea, topic, process, event, phenomenon, culture, or life story to provide a unique viewpoint or to supplement other studies

Sources of qualitative data: Words, conversations, stories, synergistic discussions, dyads, observations, documents, artifacts, reflections

Tools for qualitative data collection: Protocols, logs, diaries, journals, notebooks, rubrics, moderator guides, tools

CHAPTER 2

Using the Research Question to Guide Qualitative Data Collection Tool Design

> *Your research questions . . . are at the heart of your research design. They are the one component that directly links to all of the other components of the design . . . and will have an influence on, and should be responsive to, every part of your study.*
>
> (Maxwell, 2005, p. 65)

The Role of the Research Question in the Design of Qualitative Tools

The overview of qualitative research designs in Chapter 1 helps the reader connect the qualitative research purpose with the qualitative data collection tool. Each researcher must ask the following questions in anticipation of developing their tools: What do you want to know, and where will your data come from?

All research, regardless of the research approach, must be systematic, rigorous, and grounded in empirical data, and adhere to the following principles:

- Systematic procedures are carefully designed, with formal plans for setting up a study (an investigation or exploration) where the process is so clearly articulated and delineated that it allows others to follow the same steps for their own studies.

- Rigor is an essential element of authentic research, where the aforementioned procedures allow for corroboration and quality control and eliminate extraneous interference or undue bias.

- Empirical studies guarantee that the data can be collected, is accessible, and is based on or verifiable by observation or experience, rather than theory or assumption (Gall, Gall, & Borg, 2006).

Before any tools can be developed, the qualitative researcher must frame a research problem and purpose, identify collectible data, and clarify research questions. These questions, emanating from the problem and the purpose, determine the research design and selection of data collection tools. In the qualitative study, the choice of words is important. Words that imply quantification or measurement can be misleading; words are more meaningful in the development of qualitative instrumentation if they reflect the interpretive nature of the inquiry. For instance, using words such as *explore, uncover, discover, interpret, ascribe meaning to, assess, describe, understand,* and *perceive* relate directly to the emergent nature of qualitative research.

Constructing a qualitative purpose statement and subsequent research questions depends on answering the following questions (Moustakas, 1994; Patton, 2015; Silverman, 2013):

- What is going on here?
- What is the phenomenon under study?
- What is important in the study of this phenomenon?
- How do participants live through or experience this phenomenon?
- How do participants describe, ascribe meaning to, perceive this phenomenon?

Therefore, when developing a qualitative purpose statement, it is important to identify (1) the research design, (2) the research focus and problem, (3) participants and research site, (4) means of accessing the site, as appropriate, and (5) the conceptual framework, if applicable.

After constructing a purpose statement, research questions must be crafted that flow from the purpose statement. For qualitative designs, a central, overarching research question is typical, followed by subquestions. The central research question should include a broad question that denotes the exploration of the central phenomenon under study. The subquestions that follow these main questions are often used to probe specific aspects of the phenomenon and may also be used as the basis for interview, observation, or focus group protocols or guides.

If a conceptual framework of theory is included in the study (the operational plan for conducting the study and integrating the elements of the project), elements of the theories that help to frame the study's findings may also be included in the subquestions.

Connecting Qualitative Designs With Guiding Questions, the Research Purpose, and the Research Questions

Aligning the research design, research purpose, and research questions is a coordinated effort. Tables 2.1, 2.2, and 2.3 highlight the connections between qualitative designs, guiding questions (the overarching focus for designing a research question), purpose statements, research questions, keywords, data collection strategies, and data collection tools. Table 2.1 outlines the foundational connections between the research design, guiding questions, and the purpose statement. In this table (2.1), the purpose statement is presented as an *example*.

Table 2.1 Qualitative Research Designs, Guiding Questions, & Research Purpose Statements

Design	Guiding Question	Purpose statement (example)
Descriptive/interpretive	How can we understand a participant's experience through his or her self-constructed meaning of the phenomenon under study?	The purpose of this QL descriptive study is to describe...
Phenomenological	What is the essence of the lived experience under study?	The purpose of this phenomenological study is to uncover the lived experience of . . .

(*Continued*)

Table 2.1 (Continued)

Design	Guiding Question	Purpose statement (example)
Ethnographic	How can we study, uncover, and understand the intact culture of this group?	The purpose of this ethnographic study is to understand the culture of . . .
Narrative	What does this story(ies) reveal about this individual(s) and his or her (their) world(s)?	The purpose of this narrative study is to report the life history of . . .
Case Study	How do stakeholders describe this process or event, and what does it tell us about future practice(s)?	The purpose of this QL case study is to assess the program that contributed to the development of . . .
Grounded Theory	What theory emerges from the systematic, comparative analysis of data originating from participants sharing the same experience?	The purpose of this grounded theory study is to represent participant perspectives on the transition from . . .
Historical	How does the analysis of past events or lives of pivotal individuals inform us about the present or future state of things?	The purpose of this historical study QL is to . . .

Table 2.2 extends the elements of Table 2.1 by highlighting *sample* research questions that align with qualitative keywords, appropriate to each research design.

Table 2.2 Qualitative Research Designs, Research Questions, & Keywords

Research Design	Research Questions (example)	Keywords
Descriptive/Interpretive	How do participants construct meaning for a particular situation?	Describe, interpret, perceive
Phenomenological	How do participants describe or ascribe meaning regarding a lived experience shared by others?	Uncover, ascribe, perceive
Ethnographic	How can we understand a culture and the interactions, cultural forms, and history of an intact cultural group?	Describe, uncover, explore
Narrative	What are the stories of key individuals? What do they tell us about critical events?	Report, describe, perceive
Case Study	How can we explore a process or event that is currently underway or has already occurred that will help us understand that process or event more comprehensively?	Report, describe, assess
Grounded Theory	How can we discover the process and sequence of steps that individuals employ to adjust, change, transform, or make a transition in their lives?	Discover, describe, ascribe
Historical	How can we analyze the components of an historical event or key figure in history to understand the impact on our lives today or for the future?	Describe, analyze, contextualize

Connecting Qualitative Designs With Data Collection Strategies and Tools

Qualitative research purpose statements and research questions constitute the guideposts for determining the appropriate data collection strategies and tools for a project. While the research purpose indicates the overall intent of the study, the research questions direct the researcher to the specific data collection strategies. A detailed discussion and rationale for crafting qualitative research questions is beyond the scope of this text, but the focus here is on how to connect the purpose statement and research questions in order to help the researcher construct the appropriate tools. By designing the right tools that reflect the research objectives, qualitative researchers will obtain their data in a meaningful way. Table 2.3 builds on Tables 2.1 and 2.2 by extending the connection between research designs, data collection strategies, and the optimal data collection tools.

Table 2.3 Qualitative Research Designs, Data Collection Strategies, & Tools

Research Design	Data Collection Strategies	Data Collection Tools
Descriptive/Interpretive	Interviews, dyads Focus groups Documents Observation Reflections	Interview protocols Moderator guides Document rubrics Observation rubrics Questionnaires
Phenomenological	Depth interviews Reflections Documents	Interview protocols Journals Questionnaires Document rubrics
Ethnographic	Depth interviews Documents Artifacts Observation Reflections	Interview protocols Document rubrics Artifact rubrics Observation rubrics Discourse/ conversational tools Journals—informants Journals—researcher Field notes

Narrative	Depth interviews Reflections	Interview protocols (life history, bio) Questionnaires Journals—informant Journals—researcher
Case Study	Interviews, dyads Focus groups Documents Artifacts Observations	Interview protocols Moderator guides Document rubrics Artifact rubrics Observation rubrics Discourse/ conversational tools
Grounded Theory	Interviews Documents Reflections	Interview protocols Document rubrics Questionnaires
Historical	Interviews Documents Artifacts	Interview protocols Document rubrics Artifact rubrics

Piloting Qualitative Tools

Every data collection tool should be tested in advance of live data collection to ensure its value, integrity, and salience. If you do not conduct pretests, you run the risk of invalidating your data due to ineffective design or the inadequacy of the tool to collect data appropriately and authentically. This process is known as piloting, and it is an essential step in every study. Qualitative designs are no exception.

In most cases, piloting requires five steps: (1) identifying a pilot sample that resembles your final participant group without including your final participants, (2) preparing your tools for testing, (3) conducting the pilot test using your draft tools, (4) debriefing and assessing the effectiveness and viability of your data collection tools and procedures for using those tools, and (5) modifying your tools to reflect any changes deemed necessary and preparing them for use in your live data collection cycle. In the chapters that follow, where design is discussed, specific guidelines are offered for piloting and modifying the different types of tools. While the general pretest guidelines apply to all tools, some qualitative tools require particular strategies to ensure their effectiveness.

HIGHLIGHTS

Using the Research Question to Guide the Design of Qualitative Tools

The nature of all research consists of the use of systematic procedures and rigor and an empirical approach that includes data that can be accessed and collected.

Role of the qualitative design purpose statement and research questions: What is going on here? What is the phenomenon under study? How do particpants live through or experience that phenomenon? What is the focus of the inquiry?

Qualitative designs and guiding questions: Each specific design is guided by an overarching statement, posed as a question, that identifies the focus of that design.

Qualitative designs, guiding questions, and purpose statements determine the research questions, data collection strategies, and data collection tools that are most appropriate.

Piloting qualitative tools must follow a prescribed set of procedures before collecting live data, based on the nature of the study and the specifics of the research design/tool.

CHAPTER 3

Conducting the Qualitative Study
Researcher Role, Access, Trustworthiness, and Ethical Concerns

> [Qualitative] researchers need to consider what ethical issues might surface during the study and to plan for how these issues need to be addressed. A common misconception is that these issues only surface during data collection. They arise, however, during several phases of the research process, and they are ever expanding in scope as inquirers become more sensitive to the needs of participants, sites, stakeholders and publishers of research.
>
> (Creswell & Poth, 2018, pp. 53–54)

> The basic issue in relation to trustworthiness is simple: How can an inquirer persuade his or her audiences (including self) that the findings of an inquiry are worth paying attention to, worth taking account of? What arguments can be mounted, what criteria invoked, what questions asked, that would be persuasive on this issue?
>
> (Lincoln & Guba, 1985, p. 290)

The Researcher's Role in Qualitative Research

Qualitative researchers are often known as the instrument in a qualitative study (Denzin & Lincoln, 2003). But what does this actually mean? Is the researcher the literal repository for all data collection? Are other tools used as well, or does this statement have deeper meaning? Creswell and Poth (2018) suggest that this terminology means that a qualitative researcher is considered the primary agent of data collection, one who bridges the gap between the instrument and the data. But what of this gap?

This concept of researcher-as-instrument is actually a reflection of the distance between the researcher and the participants, documents, or other data sources. The relationship between researcher and participant is an intimate one in a qualitative study, especially in certain designs. Descriptive/interpretive designs may allow for close interactions, but they pale when compared with the relationship between researcher and informants in an ethnographic/cultural study, where the researcher may actually live with or spend considerable time with the individuals in their indigenous setting. Similarly, when conducting phenomenological studies where profound experiences or cognitive meaning-making are explored, the relationship between researcher and participant must necessarily be a close one to guarantee rich, personal data. In this way, the researcher-as-instrument does not mean that actual instruments—tools—are not used but rather that the emotional, physical, and cognitive distance between the researcher and study participants must be close enough to make that connection appear seamless. The researcher then becomes merged with the tools he or she use to collect their data. Hence, the distance between the researcher and the participant is close and personal.

The researcher's role, thus, is that of establishing working relationships with a study's participants in order to secure meaningful data for analysis. A qualitative researcher must be a skilled listener, good at probing, and patient in his or her quest for responses. He or she must be sympathetic or empathic and able to observe and interpret nonverbal cues, continuously adapting to the behaviors and responses of participants. More importantly, he or she must be able to subordinate their opinions, assumptions, and preferences in favor of their participants, so that participant perspectives drive the conversation. This reference to researcher bias leads us to position the researcher's role in a study.

Researcher Bias and the Practice of Bracketing

We are conditioned to believe that bias is a negative aspect in any research study, that bias somehow distills or dilutes the verity of the research findings. In a qualitative study, however, the impact of bias is viewed differently. While there may be a concern for undue bias in a study, that type of bias causes concern only when it diffuses the perspectives of the participants in favor of the researcher's viewpoint. A qualitative study openly and intentionally seeks the voices, perspectives, and detailed individual stories of participants, so productive bias, in the best sense, is what makes participant perspectives so valuable (Ahern, 1999; Britten, Jones, Murphy, & Stacy, 1995; Giorgi, 1994; Halquist & Musanti, 2010; Moustakas, 1994). The detrimental aspects of bias occur when researchers either neglect to disclose their own connections to or relationships with the focus of a study or when researchers let their own assumptions override the voices of their participants. One way to offset this bias is manifested through bracketing.

Bracketing is a common practice in qualitative research, grounded in phenomenology and Husserl's assertion that a researcher should suspend judgments about a research environment and participants. In this way, researchers must step back from these judgments to let the intrinsic nature of the phenomenon prevail over their interactions with participants. Although bracketing is often associated exclusively with phenomenological designs, some scholars (Creswell & Poth, 2018; Patton, 2015) apply this practice to other qualitative studies, viewing the practice as a way for researchers to defer their implicit bias in favor of the participant's viewpoint. Since all qualitative inquiry begins with the exploration of phenomenon, it makes sense to apply bracketing to other qualitative designs; if phenomenology is the basis for the qualitative approach then bracketing must, therefore, be allowed to extend beyond phenomenological designs.

In this way, therefore, researchers position themselves to make their background known, to disclose their interest in and experience with the research topic, and to acknowledge their connection in order to ensure transparency in the data collection process (Denzin, 2001; Moustakas, 1994). By creating strategies to hear participant stories, obtain rich descriptive details of participant experiences, and allow for the triangulation of multiple sources of data, the researcher allows participant views to dominate until their views surpass what the researcher *thinks* they are hearing or interpreting. Prolonged engagement in the field is one way that researchers ensure the dominance of participant perspectives; the longer you spend with someone, the more their opinions, their views, and their assertions become established as a "truth" and the less weight your own views rank in your understanding. In other words, the key is to develop a process and instrumentation that allows you, as the researcher, to obtain rich descriptive participant narratives to the extent that those narratives offset your own assumptions about what you think they mean or what you think you already know (Giorgi, 1994; Moustakas, 1994; Patton, 2002, 2015; Roulston, 2012).

The Researcher's Access to Data

Access to qualified and meaningful data depends on selecting the appropriate research site, followed by identifying those individuals who can assist in securing and maintaining access to the data.

Research Sites

The sampling strategy for all qualitative research begins with purposeful or purposive (they are considered as interchangeable) selection. Purposeful selection of participants and research sites means that the researchers select the individuals and settings for their study based on the information and insights they can provide, or, stated another way, purposeful sampling is the act of selecting information-rich cases (Patton, 2015) and criterion-based sites for your study (Flick, 2009; Maxwell, 2013; Patton, 2015). This intentional selection of individuals and locations adds to the richness of a qualitative study, where the context of the inquiry is critical to the interpretation of the findings. Carefully and purposefully selecting both participants and sites helps the researcher better understand the problem and answer the research question(s).

Gatekeepers

As formidable as this term sounds, the reality is much more relatable. Gatekeepers are individuals at a research site who provide access or links to the site, access to participants who will be included in the study, and access to other resources that facilitate the study through permissions, entry, and scheduling for the researcher. The researcher locates these individuals at each research site, often using snowball or opportunistic strategies to identify them. For instance, in a study that focuses on college presidents, a researcher may begin with the president's staff or executive assistant as a potential gatekeeper. This staff member likely has direct access to the president's schedule and e-mail correspondence and can represent his or her inclination to participate in the study. In other instances, a colleague or associate at a research site may be the first point of contact for a qualitative researcher, who then asks that individual to connect him or her with individuals who will serve as gatekeepers. This pursuit of meaningful connections is the usual way that researchers gain access and support at their chosen research locales (Marshall & Rossman, 2011; van Manen, 2014).

The researcher's responsibility to the gatekeepers is just as important as what the gatekeeper provides to the researcher (Hammersley & Atkinson, 2007; Morse, Barrett, Mayan, Olson, & Spiers, 2002; Wolcott, 2008). Typically, gatekeepers will want to know the following information as they smooth the way for the researcher's access and approval:

- Why was this site chosen for the study?
- What activities will occur at the site and in what timeframe?
- Will the study disrupt organizational activities at any time? If so, how will that disruption be managed or justified?
- How will the participants be protected, either literally or in terms of identity, privacy, and confidentiality?
- How will the results be reported? Will they be shared with the participants and others at the research site?
- What are the specific responsibilities of the gatekeeper to the researcher and vice versa?

Trustworthiness in the Qualitative Study

How do you know if a qualitative study is worthy of serious consideration? How can you be sure that the rigor and authenticity of the findings are verifiable? Researchers expend considerable effort ensuring that their studies are rigorous, valid, reliable, and actionable. Consumers expect these studies to be professionally accomplished with precision and objectivity, grounded in sound ethical practice. These expectations, however, are typically oriented toward quantitative designs; the guarantee of rigor, validity, reliability, and generalizability are not applied to qualitative designs in the same way. If different standards are applied to qualitative research to ensure quality and rigor, how can we better understand those standards to effectively evaluate qualitative studies?

Trustworthiness, a concept coined by Lincoln and Guba (1985), is considered the quintessential framework for evaluating qualitative research but receives minimal attention from most researchers, especially if they are predominantly oriented to quantitative methods. In fact, many quantitative researchers expect that the same principles of validity, reliability, and generalizability can and should be applied to qualitative designs. This is not the case, and a thorough understanding of the evaluative criteria for assessing qualitative research is a necessary component in any researcher's toolkit (Altheide & Johnson, 1994; Angen, 2000; Whittemore, Chase, & Mandle, 2001).

Four elements comprise the original trustworthiness framework: credibility (truth), dependability (consistency), transferability (applicability), and confirmability (neutrality) (Gioia, Corley, & Hamilton, 2012; Krefting, 1991; Lincoln & Guba, 1985; Patton, 1999). Authenticity is a fifth element, added since the original discussion and endorsed by some qualitative researchers as an equally important evaluative element (Polit & Beck, 2017). These elements are explained in the following section and illustrated in Table 3.1 (Billups, 2014), which shows the connection and interpretation of these elements compared with the quantitative approach. The visual display of the quantitative and qualitative terms often helps researchers understand the commonalities these evaluative criteria share.

While this chart attempts to show the progression from the quantitative to qualitative frameworks for evaluative criteria, the terms cannot be matched exactly. The following discussion attempts to further explain each element of trustworthiness.

Table 3.1 Commonalities Between Quantitative and Qualitative Research for Rigor

Quantitative Lens	Qualitative Lens	Bridge	Essential Questions
Validity	Credibility	Truth	Are results believable, seem truthful?
Reliability	Dependability	Replicable	Are results consistent over time?
Generalizability	Transferability	Applicable	Are the results applicable to similar settings?
Objectivity	Confirmability	Neutrality	Are results corroborated via triangulation?
Accuracy	Authenticity	Reality(ies)	Are all "realities" represented?

Credibility. Are the qualitative findings believable? In the quest for credible qualitative results, findings must appear truthful and capture a holistic representation of the phenomenon under exploration. The most robust of the trustworthiness strategies, there are many ways to apply credibility to evaluate qualitative findings. Five of the most common credibility strategies include prolonged engagement, peer debriefing, member-checking, triangulation, and negative case analyses.

Prolonged engagement, in conjunction with persistent observation (intense focus on the aspects of setting and phenomenon), suggests that the researcher must spend considerable time in the research field to thoroughly understand participant perspectives and to offset the researcher's own bias (although how much time depends on the nature of the study; Wallendorf & Belk, 1989). As Lincoln and Guba (1985) note, "if prolonged engagement provides scope, persistent observation provides depth" (p. 304).

Peer debriefing involves feedback from another researcher to compare conclusions; peers may address questions of bias, errors of fact, competing interpretations, convergence between data and phenomena, and the emergence of themes, all of which can be a lengthy but important process to reinforce credibility.

Member-checking differs from peer debriefing in that select participants are asked to review the findings or preliminary analysis to assess whether those findings reflect what they expressed to the researcher. This feedback may be obtained either in writing or in face-to-face conversations and is also a way to obtain supporting data, although it is primarily a data-verification strategy rather than a data collection strategy. Member checking is, however, considered controversial for many reasons and is a process that must be carefully applied (Lincoln & Guba, 1985; Polit & Beck, 2017).

Triangulation involves using multiple data sources to produce greater depth and breadth of understanding. This effort is critical in qualitative studies as a way to corroborate findings or to build a more holistic picture of the phenomenon. Triangulation is accomplished through one or more of the following approaches: methods triangulation (different data collection methods including interviews, journals, focus groups, observations, and documents), data triangulation (using different participants or data sources within one study in subsets of people, time, and space; Cohen & Crabtree, 2006), analyst triangulation (using multiple analysts to review findings through cross-case or within case analyses), or theory triangulation (using multiple theoretical perspectives to interpret the data; Lincoln & Guba, 1985).

Finally, negative case analyses include the use of disconfirming evidence to search for other interpretations in a study. The use of conflicting findings to build a richer picture of the phenomenon likewise allows for continuous refinement of the results (Creswell & Poth, 2018).

Dependability. Are the findings stable and consistent over time and across conditions? Would the same data collection methods yield the same or similar results? Ensuring that the same research process would generate the same essential findings often depends on external audits, which involve external researchers who examine the purpose, methods, and findings of a study to determine whether the findings and interpretations of one researcher can be supported by another (Miles, Huberman, & Saldana, 2014). External audits (also known as inquiry audits) are an important strategy for feedback to assess the truthfulness of preliminary findings. The drawback to this strategy, however, may instead produce conflict between two researchers' perspectives, in which case the primary researcher may decide to revise the study.

Transferability. While the goal of qualitative research is not to produce results, which are statistically generalizable, the intent is to produce findings that other researchers can interpret for similar settings, even to the point of applying the research design for their own purposes. As Trochim (2006) indicates, the concept of proximal similarity is included in this strategy. As Lincoln and Guba (1985) note, "by describing a phenomenon in sufficient detail, one can begin to evaluate the extent to which the conclusions drawn are transferable to other times, settings, situations, and people" (p. 306). This work is accomplished through the strategy of "thick description," a term first used by Ryle (1949) and extended by Geertz (1973). Thick description is the use of notes and field notes in a study where the researcher includes extensive detail and explicit descriptions when recording conversations, observations, and interpretations during data collection. Thick description allows the researcher to more easily evaluate how this same circumstance of people, place, and phenomenon could be applied in a similar setting, under similar conditions, with similar participants.

Confirmability. Are the findings accurate? Can you find other ways to corroborate your results? These efforts are crucial in a rigorous qualitative study, not only to generate confidence in the results but also to reflect the truthfulness of the participants' perspectives. There are several strategies that can be employed to apply this concept, but two of the most common include audit trails and reflexivity.

Audit trails are likened to a blueprint for the research process, outlining detailed procedural records maintained by the primary researcher. This blueprint is accessible to an external researcher, so he or she can attempt replication; if a study can be replicated with similar results, confirmability is strengthened.

Another form of confirmability is found in the practice of a researcher's reflexivity. Patton (2015) defines reflexivity as a "particular kind of reflection grounded in the in-depth, experiential, and interpersonal nature of qualitative inquiry," and involves "self-questioning and self-understanding" (p. 70). As a trustworthiness strategy, reflexive practices ensure that researchers have consciously examined what they know, how they know it, and how much of that self-knowledge affects, dilutes, or compromises what the participants have shared or expressed about their own experience.

Authenticity. This final element in the trustworthiness framework comes more recently to the discussion but is strongly endorsed by several qualitative researchers (Patton, 2015; Polit & Beck, 2017). This strategy focuses on the contextual purpose of the research, that is, what is the intended value of the research? How does the research benefit participants? Are all the realities represented to give meaning to the findings?

Arguably, the concept of trustworthiness is complex. This overview can only provide a general sense of the value and application of the elements for evaluating qualitative research. Overall, researchers who do not believe there is any way a qualitative study can be as reliable, valid, believable, or useful as a quantitative study should be assured that when the principles of trustworthiness are diligently applied, a qualitative study is just as rigorous and valuable as any quantitative study. But as Morse et al. (2002) note, "While strategies of trustworthiness may be useful in attempting to evaluate rigor, they do not in themselves ensure rigor" (p. 9). The quest for rigor in any qualitative study ultimately resides with the quality of the researcher's purpose and practice, and the verity of the unique depth and breadth of each participant's "lived experience" (Moustakas, 1994).

Ethics in Qualitative Research

Ethical Concerns

Every study that involves humans, regardless of the approach or design, must abide by the standards for the protection of human subjects, as outlined in Table 3.2. Ethical considerations in research involve the individuals' right to understand the boundaries of voluntary participation, informed consent, anonymity, privacy, or confidential treatment of their data, and the obligations of the researcher to safeguard their rights and interests. These considerations are not only safeguarded by the researcher, but they are also reviewed and enforced by institutional committees or boards who oversee the application of protections and fair treatment of all research participants at a research site.

One of the primary concerns in ethical research practices, particularly connected with qualitative research, is the establishment and evolution of the relationship between the researcher and participants. As Wang (2013) points out, the question of how the researcher jointly constructs meaning with the participants is an important issue of procedural transparency and ethical positioning. There are several ways that this positioning is affected in qualitative projects: First, the process of prolonged engagement causes researchers to shift imperceptibly from researching to researched, or put another way, they begin to see meaning as coconstructed. The distinction between the participant's views and the researcher's views become blurred; a reflexive practice, such as journaling, is one way to offset this tendency and to protect the integrity of the participant's perspectives. Second, although the researchers may bracket their position to a participant at the beginning of a study, they may need to repeat this positioning several times over the course of data collection. This renewed disclosure can be valuable not only to reassert the distinction between the researchers' interest in the study's focus but also to reaffirm their commitment to transparency and the superiority of the participant's viewpoint. Third, in instances where researchers become participant-observers or coexist with intact groups, their bias sometimes slips toward that of a participant rather than as the principal investigator. This shift can confuse the ethics of maintaining boundaries between the two roles; as Wang (2013) further notes, the researchers and the participants are motivated to engage one another during the research process, but these extended interactions may negatively affect the relationships (Elder & Miller, 1995). Becoming too close to one's participants challenges the ethical protection of informants and may strain the researcher's ability to analyze data honestly. In a similar vein, de Laine (2000) calls these inclinations "boundary violations" and cautions qualitative researchers to continuously guard against this, using reflexivity as an intervention (p. 134).

Institutional Review Boards and the Qualitative Study

As with any research project, a review of the research plan and procedures by institutional review committees is a necessary step in the process. For qualitative studies, the review process focuses on the interactions between researcher and participants, or researcher and the research site, more than on the validity of the instrument or the reliability of the data collected with that instrument. Review committees are concerned with participant rights, privacy, the confidentiality and treatment of personal and intimate data, and the procedures for informed consent in a multilayered research process. Table 3.2 outlines the information these review boards require from researchers, beginning with the plans for the research project, the procedures for conducting the study, and the assurances about how the data will be treated and reported.

Table 3.2 What Do Institutional Review Boards Want to Know About Your Study?

Planning the Study	Conducting the Study	Reporting the Findings
• Review official procedures at your research site for meeting schedules, deadlines for submissions, forms to complete	• Send informed consent forms to participants in advance, if possible	• Carefully represent all viewpoints and avoid reporting only your sole perspective on findings or the weight of a vocal minority
• Many review committees require certification by a recognized IRB organization for Principal Investigators	• Stress the voluntary nature of their roles, indicate their right to withdraw from the study without penalty at any time during the study	• Treat data carefully to protect or obscure the identity of your participants

(Continued)

Table 3.2 (Continued)

Planning the Study	Conducting the Study	Reporting the Findings
• Assess potential for risk, for the vulnerability of populations included in your study, and how your informed consent process will address these concerns	• Be clear about the results that will be shared and how they will be shared and discuss the levels of privacy, confidentiality, and anonymity	• Avoid reporting information or stories that would compromise participants or harm those at the site
• Justify your research plan and rationale to diverse audiences, and outline how you plan to approach and collaborate with gatekeepers at your research sites	• Respect your research site and participants and be careful not to disrupt or intrude on their daily work/operations	• Share results with participants and with others who have been instrumental in helping you at the research site
	• Be aware of and respect power differences at the research site; be sensitive to the cultural/societal variations in any research site and approach the participants/site with respect and deference	
	• Report any changes or variations in your study to the Review Committee, as promised	

HIGHLIGHTS

Conducting the Qualitative Study: Researcher Role, Access, Trustworthiness, and Ethical Concerns

Researchers' role in qualitative research means that researchers are considered the primary conduit between participants and the researchers due to the close interaction and intimate rapport established between them.

Bracketing is a common and essential practice in qualitative data collection to offset a researcher's biases and to allow the voice of the participant to prevail.

Researchers must purposefully select research sites for optimal data collection and must also identify individuals who can help them gain access to the site, the participants, and the resources to conduct their study.

Ethical concerns are addressed in qualitative projects by ensuring the protection of participants, their identities, privacy, and the way the data is used and reported.

IRB processes are common to most institutional sites, and qualitative projects must adhere to the requirements and procedures established by these Boards.

CHAPTER 4

Interview Protocols

> A qualitative research interview attempts to understand the world from the [participants'] point of view, to unfold the meaning of people's experiences, to uncover their lived world. . . . The main task in interviewing, therefore, is to understand the meaning of what the interviewees say. The [process] of interviewing seeks to cover both a factual and a meaning level.
>
> (Kvale, 2012, Preface and p. xvii)

Interviewing Defined

Interviewing is defined as "conversation with a purpose" (Kahn & Cannell, 1957, p. 149), while Frey and Oishi (1995) extend this definition by designating qualitative interviews as purposeful conversations in which one person asks prepared questions and the other person answers—but with detail, depth, and nuance. Kvale and Brinkmann (2014) note that "interview knowledge is produced in a conversational relation; it is contextual, linguistic, narrative, and pragmatic" (p. 21). Qualitative interviews, therefore, capture an individual's perspectives, experiences, feelings, and stories with the guidance and facilitation of an interviewer.

King and Horrocks (2011) stress the importance of the relationship between the interviewer and interviewee. Rubin and Rubin (2013) add to this emphasis by stressing the immediacy of the qualitative interview, inherent in the interactions that allow a researcher to learn about the participant's experience, by observing, listening, and gathering information that is not directly accessible. It is the interaction and the interplay between the verbal and the nonverbal, the seen and the heard, that allows for the richness of this conversational format (Alvesson, 2012). Kvale and Brinkman (2014) further add that "interview knowledge is produced in a conversational relation; it is contextual, linguistic, narrative, and pragmatic" (p. 21).

Interviewing is a time-consuming and resource intensive endeavor. Qualitative interviews are typically conducted through close contact with an individual or with individuals in a research setting. The researcher captures a holistic, integrated view of a phenomenon through the viewpoints of the participants who experience that phenomenon. Thus, interviews are an important way of allowing the participant's view of the phenomena of interest to emerge, as the participant views it and not as the researcher views it (Rubin & Rubin, 2013).

The interview process is supported by interview protocols, which are tools designed to guide, customize, and standardize the interviewing process, ensuring that the same general areas of information will be collected from each interviewee. While still allowing for flexibility and adaptability in the data collection process, an interview tool (protocol) guarantees that detailed and explicit information will be secured from the participant (Frey & Oishi, 1995; Seidman, 2013).

Interview Applications

A researcher should employ interviewing as a primary or secondary data collection strategy depending on the research design and on the goals of the research project. If the study is focused on how people think, feel, or view a phenomenon, or if personal and detailed stories are necessary to develop a narrative, then interviews are the strategy of choice. With interviews, the researcher is able to gather extensive, detailed information from fewer participants; in this way, qualitative interviewing is a rich, focused exercise compared with the broad data collection approach of quantitative research, which may attempt to capture aggregated attitudes, opinions, or perceptions through survey questionnaires or rubrics.

Many qualitative interviews are exploratory, designed to probe areas where little is known or about which new understandings are needed. They are often used in conjunction with other qualitative data collection techniques or as part of a mixed methods study. In a mixed methods project, interview data provide supplementary or enhanced findings; these data can also support the development of a quantitative instrument. Thus, interview data often lead to the development of new ideas, new insights, and new variations on phenomenon that have already been explored.

Strengths of the Qualitative Interview. Qualitative interviews offer several important advantages to the researcher:

- Useful way to obtain extensive, detailed information on a phenomenon

- Allows for data to be gathered in the participant's natural setting
- Allows for a diverse range of perspectives on the phenomenon
- Allows for immediate follow-up during the interaction
- Allows for observation as the interview progresses (i.e., nonverbal behaviors and cues)
- Can be combined with other data collection strategies to provide a holistic view of the phenomenon (observation, journaling, reflection)

Limitations of the Qualitative Interview. Conversely, qualitative interviews are limited in several ways:

- Mutual cooperation and engagement must be present for the interaction to yield useful data.
- Good interviews take time to unfold, and a lack of patience in either the participant or the researcher may negate the goals of the interview.
- Interviewees may be hesitant to be completely truthful or transparent.
- Data are subject to observer effects, the extent to which the interaction is perceived as intrusive or how reactive the researcher may be to what is disclosed.
- Interviewers may not be familiar enough with the jargon or language used by participants to comprehend their meaning.
- The interviewer's technique may compromise the quality of the data or the nature of the interaction (Brinkman & Kvale, 2014; Minichiello, Aroni, & Hays, 2008; Roulston, 2013; Rubin & Rubin, 2013).

Requisite Skills and Characteristics of the Interviewer

Effective interviewing requires skill and patience. The ability of the interviewer to show interest, put participants at ease, and gain their trust are all essential qualities to good interviewing. An interviewer who has trained and practiced in advance of conducting interviews will yield a better result. Many experts have compiled lists of the skills necessary for effective interviewing, to include content knowledge, organization, clarity, sensitivity, providing direction, critical thinking and assessment. A list of commonly referenced skills is as follows:

- Requires meticulous, thorough, careful preparation before the interview
- Must possess personal qualities of calm, quiet patience, and enthusiasm
- Must demonstrate genuine interest in what the participant is saying
- Must be able to extend compassion, empathy/sympathy, and nonjudgmental reactions
- Must be knowledgeable about the topic, be articulate in framing the questions and probes
- Must be a good listener and allow for silence as the participants frame their responses
- Must know how to respond when things are particularly sensitive or challenging
- Must possess the mental agility to operate on several different tracks, listening, covering all the questions, observing nonverbal cues, being alert to any changes in the conversation track, and knowing how to change the order of the questions as the participant responds
- Must have a good memory, remember what has been said, know which points need to be probed further (Morris, 2015; Rubin & Rubin, 2013)

There are also issues of bias implicit in the interview exchange that demand particular sensitivity on the part of the interviewer. Information gathered through any form of qualitative data collection should limit bias to the extent possible in a qualitative approach. Interviewing, however, leaves more room for bias to occur due to the intimate connection between the interviewer-interviewee. This happens in several different ways:

- As bias is inherent in the interaction between the researcher and the participant, misinterpretation is possible due to personal background and cultural differences.

- Interviewees may not express their true opinions and only tell the interviewer what they think she or he wants to hear (social desirability).

- If respondents do not clearly understand the questions, their responses may be vague or disconnected from the question.

- A researcher may convey disregard or dislike for the participant in covert or overt ways, such as disapproving of appearance, dress, or mannerisms.

- Bias can be situational, where the physical setting or location puts one or both of the participants at a disadvantage or makes either of them feel ill at ease.

- Researchers may bias the results by slanting the interpretation to their own viewpoint or what they feel their audience will want to hear (Alvesson, 2012).

What can be done to offset these tendencies or challenges? As noted earlier, Lincoln and Guba (1985) discuss the trustworthiness strategies of prolonged engagement, obtaining thick, rich description, and researcher bracketing as important ways to offset the researcher's assumptions in favor of the participant's views. Intensive listening by the interviewer; leaving space for the interviewees to tell their stories in their own ways; and using probes to generate examples, details, and nuanced descriptions are all ways to address potential bias in the interview.

Interview Formats and Types

Formats

There are several common qualitative interview formats. The predominant qualitative interview format involves a 1-to-1 interaction between the researcher and the participant. Interviews may be conducted in person, by phone, or virtually; they involve any number of strategies to connect the two individuals engaged in discourse. The traditional model, still used widely by many qualitative researchers, consists of the in-person individual interview, where a guided conversation is recorded, and the nonverbal behaviors are noted by the researcher. Other interview formats supplement interview exchanges, such as photovoice, photo elicitation, or windshield reflections (Catalani & Minkler, 2010; Epstein, Stevens, McKeever, & Baruchel, 2006; Schwandt, 2015; Wang & Burris, 1997) While photo elicitation is a data collection technique that is often used as an icebreaker to prime the participants for subsequent interviewing (Epstein et al., 2006), Wang and Burris denote photovoice as a participatory method that allows participants to use images in order to represent their worlds (1997, p. 369). Photo elicitation and photovoice strategies are akin to the interviewing process in that they allow for the participant's voice to be heard and acknowledged.

More recently, Harris (2016) suggests the benefits of the walking interview, which allows the interviewer to see the world through the interviewee's lens by sharing that world together. Defined as "interviews conducted on the move" (Clark & Emmel, 2009), this approach offers another way to conduct the 1-to-1 interview and expand the ways in which the interviewer can deepen his or her understanding of the interviewee's experience.

Another common strategy for interviewing comprises telephone interviews, which allow for some personal contact between the researcher and the interviewee, although the format does deny the researcher access to nonverbal cues (Frey & Oishi, 1995). While the researcher can still guide the conversation and hear the tone of voice or inflection, and can still clarify as the conversation unfolds, there is no opportunity to observe the connection between what the participant is saying and his or her body language or use of space. Additionally, telephone interviews need to be succinct, or it is likely that the participants will become weary or feel as if the researcher has intruded on their time. As an interview option, however, the phone interview remains a viable and often used approach.

More recently, synchronous online interviews have been conducted virtually via Skype, FaceTime, Google Hangouts, or other virtual interactive forums. While not exactly the same as an in-person interaction, the ability of the researcher and the participant to see and hear each other in real time allows for many of the same benefits as in-person interviewing.

Some researchers suggest social media, text messaging, or emailing as a vehicle for interviewing (Goodyear, Casey, & Quennerstedt, 2018; James, 2013, 2016, 2017); these strategies, while they increase accessibility and response time, and are convenient in many ways, may be limiting in their range and ability to capture depth and nuance. They are also laborious, and require precision in the case of texting and forced communication in the case of e-mailing. The back-and-forth nature of qualitative interviewing depends on the quality of the interaction between researcher and participant. The choice of format is up to the researcher and must be grounded in the research objective, but in the end, interviews can only provide rich data with the appropriate data collection tool, which consists of the interview protocol.

Types

All qualitative interviewing is denoted as depth interviewing, whereby the close interaction between researcher and interviewee generates a conversation that has an ultimate goal. There are three basic interview formats: unstructured, semistructured, and structured. A structured, standardized interview format, where a list of predetermined questions exists without much variation or follow-up, is more commonly applied in quantitative studies. By their nature, structured interviews allow for limited, forced-choice participant responses and are used mainly because of their ease of administration and ability to capture data efficiently.

For the purposes of qualitative research projects, unstructured and semistructured interview types are more common. In both instances, open-ended questions form the basis for each of these types. By open-ended, we mean that the questions are designed as nondirective inquiries that allow participants to choose their own words, context, descriptions, and meaning regarding their experiences. Spradley (1979) defines the open-ended question as an invitation; asking the participants to frame their stories by using devices such as "tell me how" or "tell me about," or even inviting the participants to describe a typical day or an example of something, thereby creating an opportunity for a rich open-ended response.

Unstructured interviews, also known as narrative, conversational, or long interviews, provide intimate access to the subjective perceptions of individuals (Doody & Noonan, 2013). Riessman (2008) and Silverman (2008) note that unstructured interviews have gained in popularity in recent years but remain challenging to facilitate effectively. In fact, DiCiccio-Bloom and Crabtree (2006) state that "no interview can truly be considered unstructured," meaning that degrees of structure are relative, and the unstructured interview is merely less proscribed and guided than the semistructured interview (p. 315).

Unstructured interviews are then defined as an interview where social interaction between the researcher and the participants informs the flow of the questions and the answers; neither are established beforehand but rather evolve as the conversation evolves (Minichiello et al., 2008). This type of interview is typically used in ethnographic studies where the flow of conversation with key informants is essential to the meaning-making of the study. Elite interviews or expert interviews are also occasions where the unstructured interview might be useful, since individuals who possess special knowledge often have a story to tell or a history to relate. In all instances, these relatively unguided dialogues are initiated by a researcher's opening prompt, after which the interviewer allows the participant to direct the flow of the conversation. The researcher does not create a prescribed set of questions for the participant; additionally, few prompts, and only prompts that are broadly directional, are used to encourage the conversation to unfold. As Punch (1986) notes, unstructured interviews allow for understanding the complex behavior of people without imposing any a priori categorization that might limit the field of inquiry. Modeled on the social conversation, the objective is to encourage important concepts to emerge, which the researcher uses to probe further. Patton (2015) offers the most succinct definition of unstructured interviewing by positing that this interview strategy "relies entirely on the spontaneous generation of questions in the natural flow of an interaction, often as part of ongoing participant observation fieldwork" (p. 437).

Semistructured interviews comprise the next level of structure in a qualitative interview. Sewell (2005) summarizes the semi-structured interview by noting its distinction from the unstructured interview; the researcher generates an outline of topics or issues that will be covered in the semistructured interview but is also free to vary the wording and sequence of the questions in order to direct and customize the interaction. While the tone of the interview remains conversational, the guidance is noticeable. There are more questions asked in this approach, and the sequence of those questions seeks to develop a storyline. Bernard, Wutich, and Ryan (2017) concur with Sewell by suggesting that semistructured interviews are flexible in that the interviewer can modify the order and details of how topics are covered, thus implying that the interviewer is in control of the process of obtaining information from the participant but is free to follow new leads as they arise. This point is critical to the success of the semistructured interview approach; the researcher must be able to adapt as the conversation unfolds in order to determine what constitutes a "lead" and to decide which questioning route should ensue.

Both of these interview strategies can be employed in a variety of research designs and with a variety of interview formats. There are instances where customized interview strategies are created for a study (see Table 4.2), but more often, a researcher uses either an unstructured or semistructured approach. Types of interviews are often associated with one or a few qualitative research designs, which require a specific type of researcher-participant interaction. Later in this chapter, Table 4.1 and Table 4.2 refer to the strategies typically employed for specific qualitative research approaches and designs.

Getting Started With a Basic Template

General Design Considerations

An interview guide, often known as a protocol, consists of the list of questions and general topics that the interviewer wishes to cover during each depth interview. Since researchers employ depth interviews to capture detailed information about an individual's beliefs, feelings, perceptions, values, and reactions to the phenomenon under study, the list of questions and topics must remain similar from interview to interview. If the list of questions and topics remains comparable, if not the same, for all the interviews, the data remain relatively consistent, allowing a story line to develop based on interviewee perspectives.

The best design approach when developing a protocol is one where the discourse follows an emergent sequence of narrative, detail, depth, and individual viewpoints. Using the predesigned protocol allows the interaction to be structured enough to cover the topics while allowing for sufficient freedom and adaptability to obtain information that might not otherwise be secured. The unstructured interview is more fluid and is less restrictive, with conversation that is open and relatively unguided by the interviewer. In the semistructured interview, probes are included to direct the conversation more carefully and to give the interviewer an opportunity to steer the narrative.

There are basic design considerations that should be included in all interview protocols.

List of topics and questions. The same list of topics and questions should be developed for all protocols used in the same study; these can vary somewhat depending on the design, and questions can be modified slightly as interviews continue. Essentially, the content should remain the same in each protocol so that data remain comparable (Josselson & Lieblich, 2015; Kvale & Brinkmann, 2009; Rubin & Rubin, 2013).

Types of question approaches. There are several basic types of topic questions that frame the way a participant constructs the descriptions of their experience:

   a. Behaviors (what a person has done/is doing)
   b. Opinions/values (what a person thinks)
   c. Feelings (intuitive versus factual)
   d. Knowledge (establishing the facts)
   e. Sensory (involving some or all of the five senses)
   f. Background (profile demographics) (Gall, Gall, & Borg, 2006).

Sequence of questions. Questions should be organized by moving from broad and less threatening or sensitive to specific and more complex. This process aids in building trust and rapport between interviewer and interviewee and allows for the story to emerge. Initial questions use wording such as "what are your thoughts about . . ." or "what are your impressions of . . ." or "tell me a bit about your experience with . . ." to develop the topics. Subsequent questions become more focused, using a technique called funneling (Patton, 2015). At this point, the use of probes become essential to delve deeply into the interviewee's perspective. As the interview concludes, questions are designed to debrief and conclude the interview, so the interviewees have a chance to summarize, enhance, or supplement their perspectives (Brinkmann & Kvale, 2014; Seidman, 2013; Rubin & Rubin, 2013).

Therefore, the general sequence of a protocol questioning route is as follows:

   a. Processual: Interviewer opens the session with an explanation of the study, purpose of the interview, and why participants were selected and provides some general sampling information, explains the background for the study, shares consent forms, and addresses protection of the individual rights, answers general questions
   b. Opening the interview session:
      i. Opening sequence: Open-ended, general questions establish trust and rapport and provide context for the more detailed questions to follow

ii. Content sequence: Specific, complex questions form the core of the interview and are written in the protocol to include probes and detailed follow-up questions to elicit thick, rich descriptive stories

iii. Concluding sequence: An effective sequence of questions is employed to allow for interviewee debriefing, summarizing, and final thoughts

Question design. Questions should be written to encourage elaboration, rather than based on dichotomous responses (yes/no). No predetermined responses are included in an unstructured or semistructured interview protocol; instead, the design should include questions that are open-ended and are supported by probes to elicit greater detail and nuance from the participant. The following guidelines apply:

a. Frame questions with reference to boundaries, that is, time, place; for instance, ask a question that specifies the timeframe in which you are exploring the participant's experience. For added richness, you can frame the question in different time dimensions, such as asking the participants how they experienced a phenomenon within the last year, then within the last five years, and so on. This dimension gives the researcher a view of how the participants have changed or viewed the phenomenon differently over time. It is recommended, however, that you ask questions about the present before you ask questions about the past or future to avoid switching timeframes on the interviewees and confusing them.

b. Avoid leading or value-laden questions; create neutral and even ambiguous questions to increase rich responses.

c. Save challenging, sensitive, or complex questions until later in your interview; create these questions as part of the second half of your protocol to give you time to establish rapport with your participants.

d. Create an interview questioning sequence that supports the conversational mode and commit to talking far less than your participants; questions should be designed to place the burden of discussion on your interviewees.

e. Effective probing is essential to produce more elaborate or rich responses. Rubin and Rubin (2013) label probes as second questions, as do other experts (Kvale, 2012; Seidman, 2013). As Patton (2015) notes (pp. 465–466), there are several types of probes that are used when designing interview protocols; the following list modifies his description by adding some of Spradley's (1979) comments on how to probe key informants in the field:

   a. Continuation probes (encourage continued talking on the present subject)
   b. Elaboration probes (asking for more detail or an explanation on a topic)
   c. Attention probes (language that lets the interviewee know that you are listening carefully, which encourages them to continue speaking)
   d. Clarification probe (asking for more detail, more explanation, or to give an example)
   e. Steering probe (guiding the interview back on track if it has lost its way)
   f. Sequence probe (establishing chronology, sequence, order, or patterns in the discussion on a topic)

Basic Interview Templates

Unstructured Interview Protocols

The key to the unstructured interview is to be more conversational than directive. In this way, unstructured interviews are conducted with minimal probing or guidance from the researcher and allow for space and silence to intervene while participants tells their "stories"; these types of interviews are typically used in narrative studies or ethnographic studies. Unstructured interviews do not consist of predetermined questions beyond the opening question; the researcher may, however, create a list of possible questions to use as probes to maintain the momentum of the interview. These interviews may take several hours to conduct and are typically employed where either significant knowledge and participant experience is required or where virtually nothing is known about the subject area. This interview type is modeled on the conversation, and—like a conversation—is seen as a social event between two people, where the rules of engagement may be mutually understood or intentionally dispensed (Corbin & Morse, 2003; Kahn & Cannell, 1957).

Develop a set of overarching themes and topics you will cover in this interview. Remember that the focus of unstructured interviews consists of the major themes you wish to explore as you participate in a conversation with this participant. The key to this type of interview is to be fluid rather than dominating, with fewer probes and more room for silence and processing—in other words, just let the participant's story unfold.

Template 4.1
Unstructured Interview Protocol

Title of Project

DATE: _____ TIME & PLACE: _____

INTERVIEWER: _____ INTERVIEWEE: _____

OTHER: _____

Pre-Interview Information & Procedures

<u>Introductions:</u> Researcher introduces himself or herself, reviews process for session, how long interview will last, and general format for questions

<u>Study purpose and applications:</u> Researcher reviews study's purpose and uses of the findings, including how the findings will be reported and shared

<u>Consent forms, approvals:</u> Informed consent forms distributed to participants, signatures secured, assurance of privacy/confidentiality/anonymity as appropriate, protection of the participant assurances reviewed, questions answered; note that the interview will be recorded and obtain permission for that, as well

<u>Treatment of data:</u> Researcher indicates how data will be managed, secured, and disposed of after a specific time period

<u>Other questions or concerns?</u> Other issues are discussed prior to beginning the interview session

Opening the Interview Session

Opening question: Use the initial question to introduce your topic and to establish a rapport with your participant.

Q1. Opening Question:

Key Interview Questions

The central portion of this interview form consists of one or two questions that set the stage for the conversational mode you are facilitating; add probes that can be used sparingly during this conversation.

Q2. Content Question:

   Probes:

Q3. Content Question (alternate or extension question):

   Probes:

Q4. Content Question (alternate or extension question):

   Probes:

Concluding the Interview

Transition to the end of your interview session with a question that allows the participants a chance to debrief or communicate any final thoughts, clarifications, or comments that still need to be shared. A single open-ended question, posed by the researcher, is the best way to capture these final sentiments or thoughts.

Q6. Concluding Question:

Researcher Script: To obtain your final thoughts, is there anything else you would like to tell me or share with me regarding today's topic?

Thank You and Follow-Up Reminder

Researcher Script: Thank you for your time and your insights on (insert topic). I will follow-up with you in a few days to (choose one or more of the following) (1) ask you to complete a reflective questionnaire, (2) complete a member-checking exercise to verify my notes of our session, or (3) ask you a few questions for clarification.

Note: An exemplar of the unstructured interview protocol is available in Appendix A.

Semistructured Interview Protocol

Semistructured interviews are the most common type of qualitative interviewing and are typically used for most qualitative designs, especially for phenomenological, case study, grounded theory, and descriptive/interpretive studies. This interview type consists of several key questions,

---

Template 4.2
Semistructured Interview Protocol

Title of Project

DATE: _____ TIME & PLACE: _____

INTERVIEWER: _____ INTERVIEWEE: _____

OTHER: _____

Pre-Interview Information & Procedures

<u>Introductions:</u> Researcher introduces himself or herself, reviews process for session, how long interview will last, and general format for questions

<u>Study purpose and applications:</u> Researcher reviews study's purpose and uses of the findings, including how the findings will be reported and shared

<u>Consent forms, approvals:</u> Informed consent forms distributed to participants, signatures secured, assurance of privacy/confidentiality/anonymity as appropriate, protection of the participant assurances reviewed, questions answered; note that the interview will be recorded and obtain permission for that, as well

<u>Treatment of data:</u> Researcher indicates how data will be managed, secured, and disposed of after a specific time period

<u>Other questions or concerns?</u> Other issues are discussed prior to beginning the interview session

Opening the Interview Session

Introductory Questions: Use these questions to introduce your topic and to establish a rapport with your participant.

Q1: Introductory Question:

Q2: Introductory Question:

Key Interview Questions

The central portion of the interview consists of questions directly related to your research question and the elements of your topic that you wish to explore. Remember to structure your questions from the broad to the specific in order to help your participant ease into the questioning route.

Q3. Content:

    Probes:

Q4. Content:

    Probes:

Q5. Content:

    Probes:

Q6. Content:

    Probes:

Concluding the Interview

Transition to the end of your interview session with one or two questions that allow the participants a chance to debrief or communicate any final thoughts, clarification, or comments that still need to be shared. A single open-ended question, posed by the researcher, is the best way to capture these final sentiments or thoughts.

Q7. Concluding Question:

Researcher script: To obtain your final thoughts, is there anything else you would like to tell me or share with me regarding today's topic?

_____

_____

Thank You and Follow-Up Reminder

Researcher Script: Thank you for your time and your insights on (insert topic). I will follow-up with you in a few days to (choose one or more of the following) (1) ask you to complete a reflective questionnaire, (2) complete a member-checking exercise to verify my notes of our session, or (3) ask you a few questions for clarification.

Note: An exemplar of the semistructured interview protocol is available in Appendix B.

organized in a specific sequence, which help the researcher define the areas to be explored but also allow for digression into related topic areas. The flexibility of this approach allows for the discovery or expansion of information important to participants previously unidentified by the researcher (Gill, Stewart, Treasure, & Chadwick, 2008).

Develop a set of questions and associated probes that you will cover in this interview. Remember that the focus of semistructured interviews centers on the key questions and the appropriate sequence of those questions. This type of interview allows for more interaction between interviewer and interviewee, and probes and guiding questions assist in the conversation.

Interviewer's Note-Taking Recording Sheet

Every qualitative interviewer uses an interview protocol to guide conversations with participants. While the protocol guides the questioning sequence or manner of asking the questions, there must also be another level of data gathering that corresponds with the interview process. The nonverbal behaviors of interviewees tell us a great deal about the truth of their statements, the level of trust and rapport that exists between the researcher and the participants, and how comfortable the participants may be with the topic.

Observing a participant during an interview is an essential part of the interviewing process, and a recording sheet is a necessary tool. While people are telling you their stories, they are also conveying what they mean and how they feel by virtue of their nonverbal and subconscious cues. Kinesics, the study of body motion, assumes that humans are constantly and unconsciously engaged in nonverbal communication. This behavior is influenced by many factors including culture, gender, age, personality, and the power distribution/exchange between the interviewer and the interviewee. Corresponding with kinesics is the study of people's space and how people behave in their settings. This focus is known as proxemics, which provides a revealing way to gather additional information about participants that either confirms or disconfirms their stories or the emphasis they place on their stories. Therefore, the interviewer must also observe and unobtrusively record things like the participants' tone of voice and speed of talking, whether they cross their arms or lean toward/away from the interviewer, whether they fidget or doodle while speaking, the position of their chest and shoulders, and the way they do or do not look directly at the interviewer.

All of these cues indicate the various levels of comfort, honesty, and rapport that add to or detract from what they are saying. For instance, if a person leans away from you and turns his or her shoulders away from you, he or she is uncomfortable and probably not being candid or sharing much useful information. Participants usually relax after a few moments, at which point their chests open up, their shoulders square off, and they will face you directly, and they may even lean toward you. These behaviors suggest increased levels of trust and rapport with the interviewer. On the other hand, people who keep their arms crossed during the entire interview or people who raise their voices or look away the entire time, suggest a level of discomfort and possibly even an attempt to obscure their real feelings about the topic. Proxemics may also come into play regarding the way a participant uses space and physical structures. If the participant is willing to sit next to you, side by side, rather than insisting on a table or desk between you, it tells you a lot about their his or her and comfort with you.

While this is not the focus of this text, the use of various devices to record nonverbal cues is important to the qualitative researcher and should be used in conjunction with interview protocols and other interview tools. The following note-taking template (4.3) provides a tool that interviewers can use to record these nonverbal cues and behaviors, either noted sporadically during the conversation or documented immediately afterward.

Template 4.3
Interviewer Note-Taking Recording Sheet

Interviewer Notes/Observations	Nonverbal Cues	Quotable Quotes

Templates Variations: Interview Protocols for Specific Qualitative Designs

As noted above, many qualitative studies employ the unstructured or the semistructured interview protocol to collect data. Table 4.1 notes these applications for specific qualitative research designs and approaches.

As Table 4.1 indicates, descriptive, elite/expert, case study, and grounded theory interviews accommodate the semistructured interview protocol format. There are, however, several qualitative designs that require customized interview protocols, which evolve from these basic templates. Three design approaches comprise this subset, outlined in Table 4.2:

- Phenomenological
- Ethnographic
- Narrative

Table 4.1 Qualitative Designs/Approaches Using Semi-Structured or Unstructured Interview Protocols

QI Design/Approaches	Interview Type & Strategy
Descriptive/Interpretive	Semistructured
Elite/Expert	Semistructured
Case Study	Semistructured
Grounded Theory	Semistructured
Historical	Unstructured, semistructured

Table 4.2 Qualitative Designs Where Customized Interview Protocols Are Recommended

QI Design	Interview Type & Strategy
Phenomenological	Phenomenological approach for "essence" meanings
Ethnographic	Ethnographic approach via key informant with cultural context
Narrative/life history	Thematic, structural, processual, dialogic, or life history/story narrative approaches

Phenomenological Interview Protocols

Phenomenological Interviewing Defined

Phenomenological interviews are designed to elicit detailed and personal stories of a group of individuals who share a common experience with a phenomenon (Moustakas, 1994; van Manen, 2014). These experiences represent a common meaning for these individuals, focusing on inherent human conditions such as joy, grief, trauma, death, transition, illness, or healing. The researcher accumulates data from these interviews and then creates an "essence meaning," which reflects a composite description of that phenomena. Phenomenological interviews are a specific type of depth interview grounded in the study of lived experiences (Arp, 2004; van Manen, 2014) in order to develop a better understanding of the ways in which people experience and make sense of their worlds.

While the disciplinary traditions of sociology, psychology, education, and a range of social and health sciences support phenomenology, phenomenological research is essentially rooted in philosophy, beginning with the work of Edmund Husserl, a 19th-century mathematician. Husserl's (1970) work serves as the foundation for phenomenological research and established the context for its later evolution. Focusing on the intentional structure of the human experience, Husserl believed that phenomenological inquiry must include individuals' emotional interpretation, physical or bodily awareness, their perceptions of self and others, and their perceptions of social and communication interactions (Arp, 2004; Pivčević, 2014). His emphasis on the consciousness of the human experience was directed at the duality of what individuals "see" as opposed to what things "mean." This distinction was protected by ensuring that the individual's experience would prevail over the researcher's interpretation or bias; therefore, Husserl proposed the concept of epoche, a suspension of one's own understanding of an experience in favor of the individual's experience (closely linked with the practice of bracketing). In this way, Husserl stressed the value of setting aside assumptions, beliefs, and interpretations of an experience by attaching meaning to that experience based on the participant's view, as the only way to preserve the integrity of the lived experience (Arp, 2004; Moustakas, 1994; van Manen, 2014).

Phenomenological Interviewing Types and Approaches

Different scholars describe different types of phenomenological approaches. Some scholars promote a longer list of phenomenology designations, such as transcendental, naturalistic, existential, generative, genetic, realistic, and hermeneutical (Creswell & Poth, 2018). These approaches could be viewed on a continuum for how objects and experiences are constituted in human consciousness and include different lenses through which to explore the "meaning of things" (van Manen, 2014). They overlap in some ways, and all of them speak to the intersection of conscious and subconscious awareness of one's self within a phenomenon.

Transcendental phenomenology, embraced by Moustakas (1994), highlights less of the researcher's interpretations and focuses more on the participant descriptions. Supporting Husserl's stance on epoche, or on setting aside personal bias in favor of the individual's descriptions, Moustakas's (1994) transcendental approach is one where the researcher combines textural descriptions of the experience (what the participant describes as tactile, visible, observable markings) with structural descriptions (the conditions, setting, situations, context). The merging of these two approaches assists in the creation of the essence meaning of the phenomenon. Many phenomenological designs reflect the transcendental approach and include the participant's descriptions of the multiple layers of an experience.

Hermeneutical phenomenology, a variation on transcendental phenomenology, comprises the study of the structures in an experience as they are experienced by individuals, where the individual provides interpretation, meaning, and justification for how they engage in their world through the phenomenon (van Manen, 2014). van Manen (2014) describes this type of inquiry as a dynamic interplay between six distinct, but overlapping, research activities: (1) Researchers identify a phenomenon of interest, (2) researchers reflect on the essential themes, (3) researchers reflect on what constitutes the nature or essence of the lived experience, (4) researchers write a description of the phenomenon, (5) researchers maintain a strong connection to the topic of inquiry and balance all the parts in relation to the whole, and (6) researchers create a composite of the phenomenon through the essence statement of the participants' lived experiences, which is their own interpretation and creation based on collective interview data (p. 86).

Defining Features of the Lived Experience and Essence Statements

The process for developing this perspective on the human experience, known as the lived experience, was refined by van Manen (2014) in what he termed the "phenomenology of practice." As van Manen explains, the purpose of phenomenological exploration is to identify the nature of the experience and to convey that lived experience to others through the detailed essence statement that allows others to understand the shared phenomenon. Many scholars suggest that what makes an experience move from subconscious to conscious, following in Husserl's original conception of the tradition, is the experience of living through something and then actualizing that experience by describing the lived experience once again (Arp, 2004; Moustakas, 1994). This process makes the phenomenological study of shared human experiences reflective of the phenomenon, broadly, and not merely representative of the individual stories (Moustakas, 1994).

Moustakas (1994) reinforces this description of "what" an individual experienced, combined with "how" they experienced it, also expressed in the researcher's creation of the essence statement. The researcher identifies, analyzes, and synthesizes the significant statements made by all participants in order to create this essence statement. This statement describes the meaning of the lived experience rooted in the unique individual viewpoints but merged to express the phenomenon holistically. Since the burden of the final synthesis rests with the researcher and his or her ability to represent these lived experiences faithfully, the phenomenological interview protocol must be designed to secure extensive details, questions that prompt the interviewees to tell their stories as if they were reliving that experience with all the moods, feelings, observations, and reflections that occurred at the time. Therefore, questions for this type of protocol must accommodate the ultimate purpose of the phenomenological approach.

Phenomenological Interview Protocol

There are several basic features in the phenomenological interview. All of these features assist the researcher/interviewer with data collection, supporting van Manen's (2014) six research writing activities that lead to the creation of the essence statement. A template design, therefore, must include sections where a researcher can record the textural (participant's first-hand statement about the experience), structural (participant's description of how he or she experienced it in terms of conditions, situations, context, setting), and the interrelatedness of self-awareness, emotion, imagination, awareness of others, intention of one's actions, memory and special recollections, communications, social interactions, and an evolving comprehension of the experience (meaning-making).

Template 4.4
Phenomenological Lived Experience Interview Protocol

Title of Project

DATE: _____ TIME & PLACE: _____

INTERVIEWER: _____ INTERVIEWEE: _____

OTHER: _____

Pre-Interview Information & Procedures

<u>Introductions:</u> Researcher introduces himself or herself, reviews process for session, how long interview will last, and general format for questions

<u>Study purpose and applications:</u> Researcher reviews study's purpose and uses of the findings, including how the findings will be reported and shared

<u>Consent forms, approvals:</u> Informed consent forms distributed to participants, signatures secured, assurance of privacy/confidentiality/anonymity as appropriate, protection of the participant assurances reviewed, questions answered; note that the interview will be recorded and obtain permission for that, as well

<u>Treatment of data:</u> Researcher indicates how data will be managed, secured, and disposed of after a specific time period

<u>Other questions or concerns?</u> Other issues are discussed prior to beginning the interview session

Opening the Interview Session

Introductory questions: Use these questions to introduce your topic and to establish a rapport with your participant.

_____

_____

Researcher Script: The purpose of this interview is to hear about your experience with (insert topic) and how you describe that experience by sharing your personal stories, insights, reactions to, and interpretation of those experiences.

Q1. Introductory Question:

Q2. Introductory Question:

Key Interview Questions

The central portion of the interview consists of questions directly related to your research question and the elements of your topic that you wish to explore. Remember to structure your questions from the broad to the specific in order to help your participant ease into the questioning route. The purpose of the phenomenological interview is to elicit significant statements about the lived experience from the participants; therefore, the questions and your probes must be directed at the deeper meaning of their experiences. Design the questions to facilitate the creation of an essence statement that reflects the shared human experience conveyed in all of the interviews

_____

_____

Q3. Content Question: (Emphasis on the overall experience and initial story about the participant's perspective)

    Probes:

Q4. Content Question: (Focus on the participants' feelings, emotions, sensations, thoughts, and observations during experience with the phenomenon)

    Probes:

Q5. Content Question: (Proceed to the textural, first-hand description of the participants' description of the experience, their awareness of and perceptions of self; their perceptions of others during that experience, for example, what were they doing and feeling? What did they perceive others were doing and feeling?)

    Probes:

Q6. Content Question: (Ask the participants to describe the structural and physical setting details such as personal location and their sense of space while in the experience, bodily awareness, other observable features and characteristics from the location)

    Probes:

(Continued)

(Continued)

Q7. Content Question: (To capture interrelatedness of individual-to-phenomenon, ask the participants to reveal or disclose critical incidents or moments during the experience that were important, transformative, and meaningful in a personal and profound way)

Probes:

Q8. Content Question: (Finally, to capture meaning-making, probe the participants to reflect on the deeper impact of the experience, on the ways in which they have since regarded or considered the experience after some distance from the experience)

Probes:

Concluding the Interview

Transition to the end of your interview session with 1 or 2 questions that allow the participants a chance to debrief or communicate any final thoughts, clarification, or comments that still need to be shared. A single open-ended question, posed by the researcher, is the best way to capture these final sentiments or thoughts.

Q9. Concluding Question:

Researcher Script: To obtain your final thoughts, is there anything else you would like to tell me or share with me regarding today's topic?

_____

_____

Thank You and Follow-Up Reminder

Researcher Script: Thank you for your time and your insights on (insert topic). I will follow-up with you in a few days to (choose one or more of the following) (1) ask you to complete a reflective questionnaire, (2) complete a member-checking exercise to verify my notes of our session, or (3) to ask you a few questions for clarification.

If the protocol template does not allow for the researcher's notes of these elements, it makes it more difficult to capture and articulate the significance and essence of the lived experience.

Ethnographic Interview Protocols

Ethnographic Interviewing Defined

Ethnographic inquiry involves the study of an intact cultural group, where the group members have lived together in the same place and for a length of time to allow for cultural norms, activities, language, and behaviors to become established and accepted. Agar (1980) described ethnography as the study of a culture-sharing group where the complex and complete description of that group's cultural fabric is detailed and analyzed.

Originating from the field of anthropology, ethnographers engage in intensive field work, searching for patterns of traditions, rituals, conventions of language, dress, and ceremony; social networks and interactions; the roles of group members; acculturation of new members; and the established rules of conduct. These observable and inferred cultural markings are uncovered through various data collection strategies, centered around interviews with culture-bearers and supported by observations recorded in field notes and the examination of cultural objects and artifacts (material culture; Atkinson, 2016; Fetterman, 2010; Stewart, 1998; Wolcott, 2008, 2009). The ultimate goal of ethnographic inquiry is to uncover and understand these observable working patterns among individuals in a group (Wolcott, 2008). In this way, ethnographies are different from other forms of cultural studies. Where culture, itself, is the focus of cultural analyses, and culture serves as the object of the inquiry, the ethnographic exploration is more concerned with the manifestation of cultural expression through the lives and interactions of those who bear and present that culture (Alvesson, 2002; Gold, 1958; Van Maanen & Barley, 1985; Wolcott, 2008).

Ethnographic Interviewing Types and Approaches

Two of the most commonly used ethnographic approaches include realist ethnography and critical ethnography. Realist ethnography is a more traditional approach, first endorsed by Van Maanen (2011) as a process where the researcher reports on the layers of cultural interactions from a third-person perspective. This accounting of the observable and discernible characteristics of an intact culture include what the researcher saw, heard, and interpreted. The major features of the daily routines and interactions of the group, the social and linguistic networks, and power systems, and the cultural markings are all recorded and converged to create a description of the culture. This approach is seen as a relatively objective characterization of an existing group, where the researcher is essentially performing as a nonparticipant observer. Interviews still serve as the primary data source in this approach, supported by observation and detailed field notes, but the storytelling remains dispassionate.

Critical ethnography is an activist approach, also known as an advocacy or transformative stance (Madison, 2012). These studies are oriented to social change intended to represent marginalized populations or populations who are unable to speak for themselves. As a design, critical ethnographies are political in orientation, with an aim to advance the needs, concerns, and rights of targeted populations. The researcher's emphasis for this type of study seeks different details, emphases, and nuances; systems of power, prestige, authority, and privilege are the focal points as the cultural description is presented (Madison, 2012).

Whether the researcher's stance is from a distance (third-person) or as an activist (first-person), the narratives are usually written in literary fashion, crafted as stories that have a structure and a logic to them (Angers & Machtmes, 2005; Hammersley, 1990). Additionally, theory often plays a role in ethnographic studies, where the researcher may begin the inquiry by applying a theory or theoretical framework to guide the exploration. These theories frame how a group's behavior might be explained and understood, such as when cultural theory, theories of acculturation, power, feminism, or conflict are applied to the project (Angrosino, 2014; de Laine, 2000; Punch, 1986).

Regardless of the emphasis or type, the approach to conducting the ethnographic study involves the same sequence of activities. The researcher must determine the culture in question, obtain access to the site through a gatekeeper and/or key informant(s), and develop an initial description of the culture. The next step is to engage in prolonged engagement with the group, studying the various aspects of group culture. The interconnections of interview data, observations, field notes, and document/artifact analysis provides the basis for developing a rich, holistic picture of the intact culture under study. As Fetterman (2010) notes, thick description is key to this type of inquiry, and data collection tools must facilitate the written records of verbatim quotes, cultural markings, social structure, power distributions, group member roles, devices of language and dress, political and religious beliefs, social relations, and the details of the group's daily lives (Fetterman, 2010, p. 125). The culture-bearer's actions form the emic perspective, while the views of the researcher comprise the etic perspective (Fetterman, 2010). Both perspectives are necessary to developing the holistic profile of the cultural workings of the group.

Defining Features of the Ethnographic Interview: The Key Informant and the Researcher's Participation

Ethnographic interviews (Fetterman, 2010; Spradley, 1979) are primarily conducted with individuals known as key informants. These individuals possess first-hand knowledge of the phenomena under study; therefore, key informant interviews seek to not only elicit participant perspectives but also seek to include contextual details of the place, time, setting, and cultural artifacts relative to the participant. These informed individuals are important linkages in an ethnographic study, helping the researcher access the inner workings of the culture. Informants are engaged based on their information-rich value and play a pivotal role within the cultural group.

Thus, key informants become the primary conduit between the researcher and the culture. The value of depending on these select and special participants is that their firsthand perspectives provide a lens through which the researcher can better understand the group without having to interview, observe, and engage with every single group member. As noted by Wolcott (2008), the three main reasons for using the special insights and information gleaned from key informants include (1) gathering needed information efficiently, (2) gaining access to that information which would otherwise be unavailable, and (3) gaining a particular interpretation of the culture from an important insider. Spradley (1979) describes the key informant as a unique connection to the cultural group under study and cautions fellow researchers to appreciate the key informant as more than just consultant, friend, respondent, or actor (p. 25, pp. 29–34). He distinguishes the variations on these different roles by identifying the nature of each type of role, but overall, he emphasizes that the key informant is willing to share information with the researcher by virtue of their special relationship.

Similarly, it is important to remember that ethnographic research involves a researcher who cohabits with that group as either an insider (participant observer) or as an outsider (nonparticipant observer). It is more typical for the researcher to participate as an insider to the group, but this largely depends on the nature of the group. A researcher who is living with a group of students in a campus sorority has a better chance of positioning herself as an insider than the researcher who studies a group of elementary children in their daily classroom interactions; in that instance, the researcher may be more likely to participate as an outsider. Further, the key to the centrality of the ethnographic researcher rests in the special relationship of the researcher with the key informants. Ethnographic interviews highlight this special connection, and protocols are designed to build on a relationship that has formed over time.

The Ethnographic Interview Protocol

Since the purpose of the ethnographic interview is to study culture from a first-hand perspective, an interview protocol must be able to capture the many levels of the culture bearer's speech, behavior, interactions, and representations of group consciousness (Fetterman, 2010; Spradley, 1979; Van Maanen, 2011; Wolcott, 2008). To this end, a range of question types frame the development of the ethnographic interview:

- Descriptive questions allow the researcher to collect samples of the respondents' language

- Structural questions allow the researcher to discover the elements of the culture in play

- Social engagement questions allow the researcher to probe social interactions, political structures, power distribution, and group member roles to elicit the ways in which the group members make their culture operational and illustrate how they live and work, day to day and year to year

- Cultural questions allow the researcher to explore how acculturation is managed and how new members become assimilated or censured or how the rules for conduct are established regarding acceptable or unacceptable behavior

- Contrast questions allow the researcher to develop meaning from the observable to the inferential

- Essence questions allow the researcher to explore details describing the group's ethos, soul, and spirit, which reflect the deeper levels of culture

Typically, ethnographic interviews are unstructured and do not include any predetermined questions, so the conversation can remain as open and fluid as possible. This orientation defers to the informant's view, priorities, and disposition; the unstructured nature of the exchange means that the interviewer must remain adaptable and follow the informant's lead. The researcher explores a few topics to help uncover the informant's meaning but lets the informant guide the conversation, the direction, the emphasis, and the outcome. Ethnographic interviews do not exist in a vacuum, however, as these data gathering activities correspond directly with fieldwork and observation, as a way to merge all the data sources into a synthesized view of a culture.

Template 4.5
Ethnographic Interview Protocol

Title of Project

DATE: _____ TIME & PLACE: _____

INTERVIEWER: _____ INTERVIEWEE: _____

OTHER: _____

Pre-Interview Information & Procedures

<u>Introductions:</u> Researcher introduces himself or herself, reviews process for session, how long interview will last, and general format for questions

<u>Study purpose and applications:</u> Researcher reviews study's purpose and uses of the findings, including how the findings will be reported and shared

<u>Consent forms, approvals:</u> Informed consent forms distributed to participants, signatures secured, assurance of privacy/confidentiality/anonymity as appropriate, protection of the participant assurances reviewed, questions answered; note that the interview will be recorded and obtain permission for that, as well

<u>Treatment of data:</u> Researcher indicates how data will be managed, secured, and disposed of after a specific time period

<u>Other questions or concerns?</u> Other issues are discussed prior to beginning the interview session

Opening the Interview Session

Introductory Questions: Use these questions to introduce your topic and to establish a rapport with your participant.

_____

_____

(Continued)

(Continued)

Researcher Script: The purpose of the ethnographic interview is to study culture from a first-hand perspective; attempt to capture the many levels of cultural markings and interactions from the participant.

Q1. Introductory Question:

Q2. Introductory Question:

Key Interview Questions

The central portion of the interview consists of questions directly related to your research question and the elements of your topic that you wish to explore. Remember to structure your questions from the broad to the specific in order to help your participant ease into the questioning route. Develop a set of content questions for the ethnographic interview by creating three types of questions used to uncover cultural data typically explored in ethnographies: descriptive (language, dress, space, objects, and artifacts), structural (knowledge about culture and context), and contrast (norms, values, beliefs, and the cultural forms that represent them). Remember that the focus of ethnographic interviews is to identify the major characteristics of the culture you wish to explore. What questions would you ask to capture the levels of culture, and who are the key informants (participants) who can provide insights and detailed information about this culture?

_____

_____

Q3. Content Question: (Emphasis on the descriptive nature of the cultural group, that is, language, dress, physical space, objects, and artifacts)

    Probes:

Q4. Content Question: (Emphasis on structural elements of the culture and its context for the cultural group under study)

Q5. Content Question: (Emphasis on how the culture was created, maintained, and led through changes and adaptation)

    Probes:

Q6. Content Question: (Emphasis on contrast and the cultural forms and markings that reflect the group's cohesion and define their interactions, such as traditions, ceremonies, rituals, saga, history, heroes/heroines, symbols, legends, myths, and other cultural forms)

    Probes:

Q7. Content Question: (Focus on how culture bearers represent the cultural forms and markings and ascribe meaning to them)

Probes:

Q8. Content Question: (Focus on the cultural interplay between culture bearers and the levels of interaction between group members, and probe how the words, gestures, and group actions are interpreted by its members)

Probes:

Concluding the Interview

Transition to the end of your interview session with one or two questions that allow the participants a chance to debrief or communicate any final thoughts, clarification, or comments that still need to be shared. A single open-ended question, posed by the researcher, is the best way to capture these final sentiments or thoughts.

Q8. Concluding Question:

Researcher Script: To obtain your final thoughts, is there anything else you would like to tell me or share with me regarding today's topic?

_____

_____

Thank You

Researcher Script: Thank you for your time and your insights on (insert topic).

Please note that follow-up is less routinized in an ethnographic study; researchers should use their judgment as to the best way to follow-up with key informants, as appropriate.

Narrative Interview Protocols

Narrative Interviewing Defined

Riessman (1993) defines narratives as "essential meaning making structures" (p. 4). Narrative interviews are defined as storytelling with a focus, where the stories of one or two individuals are collected, interpreted, and coconstructed between the individuals and the researcher

(Creswell, 2013; Creswell & Poth, 2018; Josselson & Lieblich, 2015; Riessman, 1996, 2008). These interviews not only generate a story, but they include the way the story is told by the individuals, within the context of their social, cultural, familial, linguistic, and institutional environments. The accounts of their actions or the events in which they were engaged are constructed and then chronologically represented by the researcher (Atkinson, 2016; Clandinin & Connelly, 2000). As Clandinin (2013) stresses, the story is critical, not only for what is shared or how it is told, but just as much for how the "individual's experiences were, and are, constituted, shaped, expressed, and enacted" (p. 18).

The origins of narrative research stem from disciplines such as literature, history, anthropology, sociology, sociolinguistics, psychology, and education. In all instances, the human development perspective is key to this design approach. Many fields of study have adopted their own approaches to narrative inquiry, but the narrative study is typically considered a procedure as well as a form of analysis (Creswell & Poth, 2018). Regardless of the application, narrative stories have the capacity to reveal an individual's self-perception within the context of a particular circumstance or event. Conducting the narrative interview has its own structure, but most researchers advocate for a flexible approach within this structure (Clandinin, 2013; Clandinin & Connelly, 2000; Daiute, 2013; Riessman, 2008). Creswell (2013) offers a more specific outline for developing narrative interview procedures by suggesting that the researcher selects one or two individuals who have stories or life experiences to share and then assembles the types of information that allow for the construction and coconstruction of these stories. This type of study requires multilayered data, with primary and secondary data working in tandem. Narrative interviewing is also a collaborative effort between the researcher and the participant; the question of who "owns" the story at the conclusion of data collection may sometimes present challenges when it comes to the question of who's viewpoint or version is accurate (Clandinin & Connelly, 2000; Pinnegar & Daynes, 2007).

Finally, narrative interviews are occasionally combined with another qualitative research design strategy, included as a supporting element in an ethnography or case study; narrative interviews may also be conducted as part of or as central to the historical analysis approach (Creswell & Poth, 2018; Patton, 2015). The narrative interview must, however, be interpreted in conjunction with the social, cultural, familial, vocational, and personal spheres that give meaning to those stories.

Narrative Interviewing Types and Approaches

Several scholars discuss various narrative approaches, which guide the development of several different types of interview protocols (Atkinson, 2016; Bau, 2016; Clandinin, 2013; Daiute, 2013; Gubrium & Holstein, 2013; Labov, 2006). When considering the various experts in this arena, there are three categories that can be applied to the narrative interview approach: thematic/structural/dialogic, narrative dramatism, and life history/life stories.

The thematic/structural/dialogic approach combines elements of the story that is told, how the story is told, and the performance or production aspect of the story, as in how the story evolves and conveys a message or makes a point. Riessman (2008) suggests that the individual who tells the story cannot be separated from how the story is being told, nor can those aspects be separated from what the story means in its dramatic conclusion or climax. Gee (1991) and Labov (2006) approach the narrative approach with a slightly more structural emphasis. Gee (1991) emphasizes the way the story is told over the structural progression of the story, noting the changes in tone of voice, pitch, delivery, pauses, and emphases. Labov's (2006) structural approach consists of more formal properties, where six elements must be included:

- Abstract (summary of the substance of the narrative)
- Orientation (time, place, situation, participants)
- Complicating action (sequence of events)
- Evaluation (significance and meaning of the action, attitude of the storyteller or narrator)
- Resolution (what finally happened)
- Coda (returns the perspective to the present) (p. 64)

A second category for the narrative interview is reflected in Burke's (1969) narrative dramatism approach. Burke focuses on the story as a showcase for how individuals make sense of the events in their lives. The narrative dramatism interview format uses the concept of "life as theater" to interview participants and follow a construct (five sequenced steps) to reveal how participants experienced or describe a particular phenomenon.

Burke's (1969) five key elements in any individual's story include the following:

- Act: What took place?
- Scene: What is the backdrop of the act or situation that occurred?
- Agent: Who is the person who performed the act?
- Agency: What means or props did the agent use?
- Purpose: What is the "why" behind the act? (pp. ix–x)

These five elements do not stand on their own but are meant to be employed as paired elements. These pairings, or ratios, provide different ways for readers to understand the story as it unfolds. Act and Agent might be paired, just as Agent and Scene might be paired; these pairings create a multidimensional view of the individual's experience, and they reinforce the contextual framework so important to the narrative's development. The researcher constructs a "script" of the participant's experience by intersecting two or more of the dramatic conventions to showcase the interconnectedness of the parts of the experience and how they provide insight into the story (Brock, 1985; Burke, 1969).

A third category for the narrative is found in the life history or life story interview. A researcher orders the meaning of individual experiences in the life stages sequence, employing a three-phased process: past life history, present life accounts, and the integrated meaning of the two perspectives. Seidman (2013) proposes a model that offers one approach to the life history interview. His three-phase process occurs emergently, with each interview spaced between three days to one week apart. This process allows a researcher to begin with a life story/history, then capture the details of that life experience, and conclude with the reflection of what those experiences meant to the individual. The life history interview format can be easily adapted to a narrative interview protocol (pp. 20–23).

Defining Features of the Narrative Interview

The narrative approach comprises a specific set of features that enable the researcher to collect the stories of individuals in a comprehensive and intensive manner. As Daiute (2013) notes, the organizational structures for recounting a story must be done so that the story becomes vivid and relatable for the reader. Therefore, regardless of the type, the narrative interview protocol must reflect multiple levels of storytelling, story-hearing, and story-recording (Miller, 2015).

Narrative projects include the following design features:

a. The focus of the narrative exploration is about the story, either spoken or written, or both.

b. The stories revolve around specific places, situations, and circumstances, making the contextual details an essential piece of the story.

c. Stories are supported by numerous types of data, including observations and field notes, participant journals/diaries, researcher journals/reflexivity, extant documents and artifacts, stories and observations by family members or other members of the individual's social network, written correspondence exchanged between the participant and other relevant individuals, and other personal-social artifacts.

d. The researcher shapes the data into a chronological storyline, situating the stories within the participant's cultural, social, familial, and professional context, or the researcher shapes the data into a theatrical or production saga.

e. Narrative interviews employ a collaborative connection between researcher and participant, where the story becomes coconstructed (Riessman, 2008).

f. Various analytical strategies may be used to restory the stories (Cortazzi, 1993; Creswell & Poth, 2018; Kvale, 2012) by designating a beginning, middle, and ending, with protagonists and antagonists, and a predicament and climax, as features of good storytelling.

g. Similarly, narrative stories reflect turning points, crossroad incidents, and epiphany moments in the participant's life (Clandinin, 2013; Lypka, 2017; Pennebaker & Seagal, 1999; Pinnegar & Daynes, 2007).

The Narrative Interview Protocols

For the purposes of developing narrative interview protocols, there are three broad categories of type: narrative thematic/structural/dialogic, narrative dramatism, and narrative life history/life stories. These types are outlined in the template variations that follow (4.6, 4.7, 4.8).

Template 4.6
Narrative Thematic/Structural/Dialogic Interview Protocol

The purpose of the thematic/structural/dialogic narrative interview is to elicit the participants' perspective on the story as they experienced it, as well as how they tell the story, the production aspect of the storytelling, such as how the story evolves; the structural, textural, and emotional elements of the story; and the details in the story that lead to the message that the story conveys or represents.

Title of Project

DATE: _____ TIME & PLACE: _____

INTERVIEWER: _____ INTERVIEWEE: _____

OTHER: _____

Pre-Interview Information & Procedures

<u>Introductions:</u> Researcher introduces himself or herself, reviews process for session, how long interview will last, and general format for questions

<u>Study purpose and applications:</u> Researcher reviews study's purpose and uses of the findings, including how the findings will be reported and shared

<u>Consent forms, approvals:</u> Informed consent forms distributed to participants, signatures secured, assurance of privacy/confidentiality/anonymity as appropriate, protection of the participant assurances reviewed, questions answered; note that the interview will be recorded and obtain permission for that, as well

<u>Treatment of data:</u> Researcher indicates how data will be managed, secured, and disposed of after a specific time period

<u>Other questions or concerns?</u> Other issues are discussed prior to beginning the interview session

Opening the Interview Session

Introductory Questions: Establish the setting and context for the narrative that the participant will share with you and introduce the levels of questioning you will ask.

_____

_____

Researcher Script: The purpose of this interview is to ask you to tell me the story about (insert topic) and to provide as many details about the story as you can recall; we will begin with some general questions about the highlights of the story and the players and the setting and then delve into particular details about each aspect of your story.

Q1. Introductory Question: (General opening questions to invite the participant to provide the basic elements of the story and set the scene for later details and examples)

(Researcher notes the tone of voice, pitch, and emphasis on certain aspects as the participant shares the story through the entire interview)

Key Interview Questions

The central portion of the interview consists of questions directly related to your research question and the elements of your topic that you wish to explore. Remember to structure your questions from the broad to the specific in order to help your participant ease into the questioning route.

_____

_____

Q2. Content Question: (Orientation of time, place, situation, context, and participants involved in the story)

  Probes:

Researcher notes on tone, pitch, emphasis:

Q3. Content Question: (Focus on the participants' details as they build the action of the story, identify key players, key moments, leading toward story's climax)

  Probes:

Researcher notes on tone, pitch, emphasis:

(Continued)

(Continued)

Q4. Content Question: (Ask the participant to detail the sequence of events, meaning of the actions, and the story's progression to the climax, describing the climax in full detail)

Probes:

Researcher notes on tone, pitch, emphasis:

Q5. Content Question: (Ask the participant to describe how the storyline was resolved and what happened to key players, how issues were addressed/discussed, and what the resolution meant to the essence of the story)

Probes:

Researcher notes on tone, pitch, emphasis:

Concluding the Interview

Transition to the end of your interview session with one or two questions that allow the participant a chance to debrief or communicate any final thoughts, clarification, or comments that still need to be shared. A single open-ended question, posed by the researcher, is the best way to capture these final sentiments or thoughts.

Q6. Concluding Question: (Coda—coming back to the present, meaning-making for the participant, asking them how that story feels to them now, as they have reflected on it and shared it)

Researcher Script: To obtain your final thoughts, is there anything else you would like to tell me or share with me regarding today's topic?

Thank You and Follow-Up Reminder

Researcher Script: Thank you for your time and your insights on (insert topic). I will follow-up with you in a few days to (choose one or more of the following) (1) ask you to complete a reflective questionnaire, (2) complete a member-checking exercise to verify my notes of our session, or (3) ask you a few questions for clarification.

Template 4.7
Narrative Dramatism Interview Protocol

The purpose of the narrative dramatism interview is to invite participants to tell their stories as a form of theater or as a dramaturgical production. Researchers work with their participants to help them make sense of the events in their lives by using drama as the vehicle by employing five sequenced steps to explore the story's aspects: act, scene, agent, agency, and purpose.

Title of Project

DATE: _____ TIME & PLACE: _____

INTERVIEWER: _____ INTERVIEWEE: _____

OTHER: _____

Pre-Interview Information & Procedures

Introductions: Researcher introduces himself or herself, reviews process for session, how long interview will last, and general format for questions

Study purpose and applications: Researcher reviews study's purpose and uses of the findings, including how the findings will be reported and shared

Consent forms, approvals: Informed consent forms distributed to participants, signatures secured, assurance of privacy/confidentiality/anonymity as appropriate, protection of the participant assurances reviewed, questions answered; note that the interview will be recorded and obtain permission for that, as well

Treatment of data: Researcher indicates how data will be managed, secured, and disposed of after a specific time period

Other questions or concerns? Other issues are discussed prior to beginning the interview session

Opening the Interview Session

Introductory questions: Use these questions to introduce your topic and to establish a rapport with your participant.

_____

_____

(Continued)

(Continued)

Researcher Script: The purpose of this interview is to hear about your experience with (insert topic) and how you describe that experience by sharing your personal stories, insights, reactions to, and interpretation of those experiences.

Q1. Introductory Question: Setting the stage (Solicit a synopsis of the story from the participants to establish the basic components of the drama and to frame the details to follow, such as the key actors, the scene/setting, storyline, and supporting roles/props).

Key Interview Questions

The central portion of the interview consists of questions directly related to your research question and the elements of your topic that you wish to explore. Remember to structure your questions from the broad to the specific in order to help your participants ease into the questioning route.

Q2. Content Question: Act (What took place and when? What are the details about how the story unfolded, and what were the major turning points in the story?)

   Probes:

Q3. Content Question: Scene (Ask about the setting, locale, backdrop, scenery, and any other context issues that frame the action)

   Probes:

Q4. Content Question: Agent (Where were the primary agents of the drama, who were the main actors and supporting actors? Who drove the action as the protagonist? Antagonist?)

   Probes:

Q5. Content Question: Agency (Ask about the props, contextual elements that help to propel drama)

   Probes:

Q6. Content Question: Purpose (Ask about the rationale, purpose of the drama, reason for the action)

   Probes:

Concluding the Interview

Transition to the end of your interview session with one or two questions that allow the participants a chance to debrief or communicate any final thoughts, clarification, or comments that still need to be shared. A single open-ended question, posed by the researcher, is the best way to capture these final sentiments or thoughts.

Q7. Concluding Question: Synthesis of elements for pairing (bring all the elements together in a participant-focused epilogue to facilitate the various combinations of the pedantic elements).

Researcher Script: To obtain your final thoughts, is there anything else you would like to tell me or share with me regarding today's topic?

___

___

Thank You and Follow-Up Reminder

Researcher Script: Thank you for your time and your insights on (insert topic). I will follow-up with you in a few days to (choose one or more of the following) (1) ask you to complete a reflective questionnaire, (2) complete a member-checking exercise to verify my notes of our session, or (3) ask you a few questions for clarification.

Template 4.8
Narrative Life History/Life Story Interview Protocol

The purpose of the life history/life story interview is to work with a participant to coconstruct the meaning of individual experiences in the stages of the participant's life, employing a three-phased process: past life history, present life accounts, and crossroad interpretations of the past life and the integrated meaning of the two perspectives (Miller, 2015; Seidman, 2013). As Seidman (2013) further notes, the "spacing between the three phases should span between three days to one week to ensure the best results"; this delay between interview sessions allows a participant to reflect, process, and enhance the next phase of the interview (pp. 20–23).

(Continued)

(Continued)

Title of Project

DATE: _____ TIME & PLACE: _____

INTERVIEWER: _____ INTERVIEWEE: _____

OTHER: _____

Pre-Interview Information & Procedures

<u>Introductions:</u> Researcher introduces himself or herself, reviews process for session, how long interview will last, and general format for questions

<u>Study purpose and applications:</u> Researcher reviews study's purpose and uses of the findings, including how the findings will be reported and shared

<u>Consent forms, approvals:</u> Informed consent forms distributed to participant, signatures secured, assurance of privacy/confidentiality/anonymity as appropriate, protection of the participant assurances reviewed, questions answered; note that the interview will be recorded and obtain permission for that, as well

<u>Treatment of data:</u> Researcher indicates how data will be managed, secured, and disposed of after a specific time period

<u>Other questions or concerns?</u> Other issues are discussed prior to beginning the interview session

Opening the Interview Session

Introductory Questions: Use these questions to introduce your topic and to establish a rapport with your participant.

_____

_____

Researcher Script: The purpose of this interview is to hear about your experience with (insert topic) and how you describe that experience by sharing your personal stories, insights, reactions to, and interpretation of those experiences.

Q1. Introductory Question: (Begin the discussion by asking basic demographic and profile information about the participants, their backgrounds, and two to four themes or phrases they would use to describe their lives; this positions the participants to frame their focus, their perspective, and to begin to organize their thoughts about how to tell the story of their lives).

Key Interview Questions

The central portion of the interview consists of questions directly related to your research question and the elements of your topic that you wish to explore. Remember to structure your questions from the broad to the specific in order to help your participants ease into the questioning route.

_____

_____

Q2. Content Question: Past life accounts (Open the interview with one of two different questions that encompass a reflection of one's life: Ask the participants to reflect on their lives by identifying three to five significant turning points in their lives, or ask the participants to reflect on their life at five-year intervals, beginning at any point they choose)

    Probes:

Q3. Content Question: Present life accounts and crossroad moments (Ask the participants to reveal the details of the life story they are reconstructing and to outline the details of their lives by giving examples)

    Probes:

Q4. Content Question: Integration (Probe the participants to reflect on the deeper impact and transformative nature of their experiences, on their crossroad moments, on the ways in which they have since regarded or considered the experience after some distance from the experience, and how they view their lives today in light of those experiences)

    Probes:

Concluding the Interview

Transition to the end of your interview session with one or two questions that allow the participants a chance to debrief or communicate any final thoughts, clarification, or comments that still need to be shared. A single open-ended question, posed by the researcher, is the best way to capture these final sentiments or thoughts.

Q5. Concluding Question: Life + World (Coconstruct the story with the participants by reviewing the stories they have shared and their context in terms of their social, physical, cultural, familial, and professional/vocational experiences)

(Continued)

(Continued)

Researcher Script: To obtain your final thoughts, is there anything else you would like to tell me or share with me regarding today's topic?

_____

_____

Thank You and Follow-Up Reminder

Researcher Script: Thank you for your time and your insights on (insert topic). I will follow-up with you in a few days to (choose one or more of the following) (1) ask you to complete a reflective questionnaire, (2) complete a member-checking exercise to verify my notes of our session, or (3) ask you a few questions for clarification.

Piloting the Interview Protocol

As with all research projects, it is important to test your data collection tools prior to collecting live data. Piloting or pretesting is an essential step in your procedures and will ensure that your tools are viable, that your process for engaging your participants is workable, and that your instructions for the interview are clear and understandable. All of the interview protocols discussed in this chapter should follow the same procedures for piloting.

There are three steps to conducting a pilot for your interview protocols, and they include preparing your interview protocols for piloting, conducting the pilot interviews, and postpilot assessment of the protocols and the process.

First, a preliminary draft of your interview protocol should be reviewed thoroughly before testing it with a pilot participant. It is important to rehearse the interviewing process, and a pilot also provides you with a chance to organize your materials and recording equipment. Review your protocol language and instructions carefully before trying it out; make sure you have the consent forms, any supporting materials you wish to provide to your participants, and any incentives you have promised.

Second, invite one or two individuals who mirror but who will not be included in your final sample to pretest your tools. Once you have selected participants who will help you, outline a list of questions you wish to ask them at the conclusion of the test. For instance, you will want to test the clarity of the protocol questions, the sequencing of the questions, the use of terms or industry-specific language, the timing of the process, and your ability as an interviewer to probe and capture rich details and stories from the participants.

Choose a setting that is free from distraction and is relatively neutral for both parties. Run through all the phases of the interview process, starting with an explanation of the study's purpose, the informed consent process and signing of forms, the rights of the participant, and the guarantee of confidentiality (or even privacy). Explain the format of the interview, and indicate how long the interview will take. Allow for a few general questions about the process and the study before you begin.

As you begin to ask your questions, monitor your own skills and inclinations as an interviewer. Make eye contact, and avoid spending the entire time taking notes; be patient as your interviewees think about their answers. Make every effort to keep the discussion going and stay on topic while leaving time for the participants to take the conversation where it may need to go. Be alert to the need for probes to secure details, examples, and stories from each participant's perspective; make notes about where your interview questions might need refinement, where you might need additional probes, and where you have redundant questions. Whenever it seems like an interviewee is replying in general terms or using language that is too broad, use probes to bring the discussion back to a more detailed, nuanced level.

Listen actively; try not to interrupt your interviewees or employ facial or body gestures that communicate your approval or disapproval regarding their statements. This takes considerable control, but it is important to stay as neutral as possible in order to allow the participant's story to emerge. When you ask questions, keep them simple and brief; when you provide probes, offer open-ended probes that encourage interviewees to provide extensive and additional details about their experiences. Only with practice can you perfect the skills necessary to conduct a quality interview; the pilot exercise is an important way to perfect these skills.

Occasionally test your audio recording equipment; it is actually best to bring two recording devices in case one of them fails. Also, make sure to observe the nonverbal language of your interviewees to watch for distraction, boredom, fatigue, or confusion; be alert so you do not lose control of the interview.

Finally, as you conclude the pilot interview, ask your participants for feedback on the questions, sequence, and tone of the interview. Ask them to offer insights and recommendations about areas needing improvement. Compile a list of notes and feedback to apply to your interview protocol for final application in your live data collection phase.

Transforming Interview Data for Analysis

Interview data are typically recorded (audio, video) and transformed into a file that can be e-mailed or downloaded for transcription/production by any number of companies that provide this service for a fee. Using high quality recording tools that allow for the creation of audio files to e-mail or download to a third-party service is a critical piece of your data management practice and will guarantee that your interview data are appropriately prepared for data analysis. Likewise, it is important to select a company that is known for reproducing transcripts of high quality, verifiable conversions of the interviews into usable narratives, all within a reasonable amount of turnaround time and for reasonable costs.

Transcripts are the essence of analysis for interview data; without a readable, understandable transcript of your interviews, you have nothing to analyze. Depending on the data analysis strategy you have selected for your study, your transcripts will be used in various ways. Overall, however, you will need to manage your data through the process of data cleaning, data reduction, and data coding. After those initial stages of data management, the analysis strategies may lead you to different approaches and uses of the transcript data.

Despite the seemingly endless resources regarding qualitative research, there are few extant standardized rules or procedures for managing qualitative data; furthermore, the process is labor intensive and largely intuitive. Interview data, in particular, require a careful procedural plan in order to extract the meanings, themes, and interpretations that are essential to answering your research questions. The combination of continuous analysis and intensive immersion with your data can make this process seem overwhelming. Experts such as Bernard et al. (2017), Boeije (2010), Grbich (2012), and Miles and Huberman (1994) discuss the myriad data analysis options. Overall, however, there are a few basic steps that all researchers can employ to successfully manage their qualitative data that are common to all research designs.

While this brief overview is not designed to replace a more in-depth discussion on qualitative data analysis, I would like to offer a blueprint for beginning qualitative researchers to help you develop a better understanding of the process. There are essentially four basic steps involved in all qualitative data management, and they are particularly important when preparing interview data for analysis:

1. Phase I: Raw data management (working with the words and notes from transcriptions

2. Phase II: Data reduction (the process of selecting, focusing, simplifying, abstracting, and transforming raw data into workable "chunks" or categories)

3. Phase III: Data analysis and interpretation (the process of analyzing data to tell the story, represent the experience, reflect the essence of the participants' perspective)

4. Phase IV: Data representation (the process of compressing an array of information into an organized pattern of findings that allows for conclusions and recommendations) (Billups, 2012)

For the purposes of this chapter, Phases I and II will provide a summary of how you can prepare your interview data for the analysis strategy that matches your research design. In Phase I, you should approach your raw data management by reviewing your transcripts and notes from the interviews; these are your raw data. You may begin with a few to hundreds of pages of transcripts or notes; in many instances, the beginning researcher may not know where to start. You must "clean" or prepare your raw qualitative data, presented in the form of interview transcripts. Except for the grounded theory analytical approach (constant comparative), your best starting point is to immerse yourself in your data by reading your notes holistically. Over the course of a few days, reading your notes or transcripts as an entity and making notes in the margins (often called memoing) will help you begin to understand, internalize, and make sense of your data. This immersion process prepares you for Phase II.

In Phase II, known as the data reduction phase, you will develop a preliminary set of codes or categories that you use to cluster the raw data into units or chunks that share similar qualities (this is often called winnowing). The data reduction process involves four distinct steps: (1) initial coding where you create preliminary codes that are either a priori (meaning they are pre-existent codes derived from theoretical frameworks or the literature) or in vivo codes (derived from the raw data, i.e., emerging from participants' words, using context-bound jargon or language); (2) secondary coding (developing a code book, revising and consolidating codes, and labeling final code categories—codes are often referred to as pattern/descriptive/interpretative; Miles & Huberman, 1994), first cycle/second cycle coding (Miles, Huberman, & Saldana, 2014), or open/axial/selective for grounded theory analysis (Birks & Mills, 2011; Corbin & Strauss, 2015); (3) clustering (assigning groups of related, coded data into clusters and assigning preliminary labels to those clusters that will eventually become preliminary thematic or content labels); and (4) thematic groupings (organizing clusters of codes into groups that generate meaningful themes, which relate back to participant words and allow for meaning to be assigned to themes, and you may use a participants' words as the labels for these groupings).

As you can see, you begin with a large volume of words/phrases/notes, and you must work to reduce and collapse these data into clusters that share similar meanings; from these clusters, you generate themes. While you may begin with as many as 30 to 40 initial raw code categories, you will eventually reduce your code groups until you end up with a smaller number. You may find that some codes can be assigned to multiple clusters—this is a very common occurrence; you may also discard some of your coded data, finding that they can be consolidated into another set of data or that they are not meaningful. All of these decisions are part of the data reduction phase.

At this point in the process, data reduction involves a variety of computer assistance, visual aids, or hands-on tactics. Many researchers use large Post-It note wall pads to create initial and subsequent codes and clusters; visualizing the data on a large surface, like a wall or board, allows for synthesis and understanding. Other researchers use the "long table" approach (Krueger & Casey, 2015) to organize data with pieces of paper spread out over a table; labeling transcripts with colored markers and cutting the pages into strips, organized by highlighter color (coding) is another way to see the codes and categories take shape.

Word processing tools can also be used to organize data into subunits. While this overview does not intend to include a detailed discussion of the many computer software programs that facilitate data management, some of the more common packages include ATLAS/TI, HyperRESEARCH, NVivo, MaxQDA, or NUD*IST. These packages assist with organization, clustering, concept mapping, and even theory development (Kuckartz, 2014; Miles et al., 2014).

The third and fourth stages of data management and analysis move toward specific strategies designated for the qualitative design you have employed. The process in this phase cannot be entirely separated from the data reduction phase, as it involves a continuous review of the data as the codes and clusters are developed. The themes that emerge from the data become the story or the narrative. Generally, the process of organizing, coding, recoding, and creating thematic categories allows you to see the emergent concepts that tell a story.

The analytic strategies applied to specific designs are varied. For instance, if you conduct a descriptive or interpretative design, you may apply Boyatzis's thematic strategy (1998) or Colaizzi's (1978); if you conduct a phenomenological study, you may use Moustakas's (1994) or Giorgi's (1994) holistic or "essence meaning" approach. If you used a grounded theory design, you will use the constant comparative analytical approach to develop theory (Corbin & Strauss, 2015; Glaser, 1993). Yussen and Ozcan (1996) propose a five-step approach that likens storytelling to a play or theatrical portrayal, which is a strategy applicable to life story/life history interview data. Sullivan (2013) outlines a dialogical approach to analyzing data to craft participants' stories. You must decide which approach best suits your study and what you seek to discover. While there are many viable options for analysis, Miles et al. (2014), Creswell and Poth (2018), Bernard et al. (2017), and Boeije (2010) provide excellent summaries of some of the more common analysis methods, which serve as a useful starting point for the beginner.

Regardless of your final selection of an analysis strategy, you must be sufficiently immersed in your data, not only to accurately represent your findings but to also communicate alternative meanings and help your readers feel as if they were living the participants' experience. The end result of this process will have subsequently transformed pages of raw notes into a meaningful narrative, representing the voices and perspectives of your participants (Pennebaker & Seagal, 1999).

HIGHLIGHTS

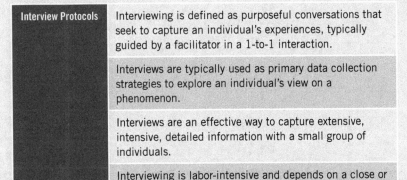

Interview Protocols	Interviewing is defined as purposeful conversations that seek to capture an individual's experiences, typically guided by a facilitator in a 1-to-1 interaction.
	Interviews are typically used as primary data collection strategies to explore an individual's view on a phenomenon.
	Interviews are an effective way to capture extensive, intensive, detailed information with a small group of individuals.
	Interviewing is labor-intensive and depends on a close or productive relationship with the interviewee.

(Continued)

(Continued)

> Qualitative interviewers require skills of patience, active listening, genuine interest in the person's experiences, and an ability to respond and probe sensitively.

> Bias in interviewing is inherent and can be moderated by bracketing and personal disclosures with the interviewee.

> Interviews are typically conducted in 1-to-1 formats in person, on the phone, through social media, via photographic or recording tools, or virtually.

> Interviews often consist of either unstructured or semistructured formats; each qualitative research design may also use a specific format or approach in its design.

> Interview guides are known as protocols and follow a standard sequence of questions ranging from introductory questions to content questions to closing questions; specific designs apply specific features for the protocols.

> Interview protocols must be pretested in pilot studies to test the instrument for live data collection.

> Interview data are usually recorded and transcribed so that they can be prepared for data analysis, using any number of analysis strategies that match the research design.

CHAPTER 5

Conversational and Discourse Analysis Tools

> *Conversational and discourse analysis relies on naturally occurring interactions and conversations, close to the phenomenon, in an emergent event where the purpose is to analyze how people act as they do, manifested through their "talk" and their behaviors while they "talk."*
>
> (ten Have, 2007, pp. 9–10)

Conversational and Discourse Analysis Defined

Conversational analysis (CA) and discourse analysis (DA) consist of the "in situ" study of ongoing dialogue among individuals or groups, where the focus is on understanding how people interact while they are talking. Sacks, Schegloff, and Jefferson (1974) first described these forms as the ongoing interpretation of how participants interact within the underlying social frameworks in which they live. Sacks et al. further stressed the procedural analysis of verbal interactions that revealed how individuals solved organizational problems, how they operated within and across groups and group norms, and how they leveraged one conversation into another. Considered by some researchers as a form of natural, in-process, everyday conversation (Bischoping & Gazso, 2015; Clandinin, 2013; Clandinin & Connelly, 2000; Gubrium & Holstein, 2013), conversational analysis and discourse analysis is essentially about "analyzing talk" (Wooffitt, 2014). The approach allows researchers to better understand the structure, patterns, organization, and routines of participants from an insider's perspective (Cohen & Crabtree, 2006; Lester & O'Reilly, 2018).

Many scholars view CA as something separate from DA. Conversation analysis is defined as the study of the structure and meaning of talking, with the consideration of how language is used by individuals. The examination of pairing and sequences of language, the nonverbal behaviors that accompany the conversation, and the roles those individuals assume as they converse (Livholts & Tamboukou, 2015; Lypka, 2017; Wooffitt, 2014) comprise the essence of CA. Originally rooted in the sociological tradition, CA was quickly adopted by psychologists, anthropologists, and linguists (ten Have, 2012).

DA is more difficult to define. Scholars define and apply DA in different ways (Bischoping & Gazso, 2015; Clandinin, 2013; ten Have, 2012); the psychological perspective frames discourse as a form of discursive psychology, or the uncovering of mental states (Potter & Wetherell, 2005; Reissman, 1993). Some scholars view DA as an extension of CA, where researchers can examine how power and influence shapes the interplay between individuals in the midst of conversation (Lypka, 2017). More specifically, Rogers (2004) distinguishes critical discourse analysis as a form of DA, emphasizing the study of the relationships between language, form, and function, which helps researchers understand how participants create meaning from socially defined or bounded practices. Regardless of the similarities in focus, critical discourse is viewed as a subset of the broader category of DA for the purpose of this discussion.

There is minimal agreement among scholars about the absolute boundaries that separate conversational and discourse analysis as data collection strategies. More often than not, conversational and discourse analysis are seen as representing interval points along a continuum. In any form, these data may be available in the spoken conversation (heard, observed) or in the written word (journal entries, reflections, prose); both formats are acceptable for analysis.

While better understood within the frameworks of ethnomethodological perspectives, these data collection strategies offer many advantages to the qualitative researcher, bridging the researcher-directed conversation in an interview or focus group, with the researcher-passive role of observation or third-party review of participant-created journal or diary entries (Potter & Wetherell, 2005; Tseliou, 2013; Wiggins, 2017). Conversational and discourse analysis strategies provide an opportunity for the researcher to follow the participant's line of thinking and pattern of conversation as a socially constructed activity. As Lypka (2017) reflects on the work of Bischoping and Gazso (2015), she stresses the importance of the researcher's position, subjectivity, and reflexivity in the gathering and analysis of these data. As she further suggests, to truly understand conversations among individuals, researchers must position themselves close enough to understand but sufficiently removed to remain open-minded.

Conversational and Discourse Analysis Applications

Conversational and discourse data frequently serve as supporting or supplementary data in a study where the primary data sources are already conversational or interactive in nature. These data are positioned in the context of dialogue that precedes or follows the conversations under study (Heritage, 1988). These data also serve as triangulation data, corroborating findings with primary sources and other secondary sources, such as document and artifact analysis, observation, and reflections (van Lier, 1988). Ethnographic and cultural studies benefit from these data as a reference for other participant dialogue and reflections and to frame researcher observations of the participant interactions in the natural setting (Fairlough, 2010; Gee, 2014a; Lypka, 2017). Grounded theory studies may also benefit from these data, where studying the structure and meaning of conversation provides supplementary data to expand the focus on an individual's transition or adjustment to a process or event (Lypka, 2017; Rapley, 2011).

Requisite Skills and Characteristics of the Conversational and Discourse Analysis Researcher

Like any type of qualitative engagement, researchers must engage in a sensitive and self-aware positioning of themselves as they participate in conversational or discourse analysis activities. Researchers who possess skills of unobtrusive listening, openness to the content and delivery of the conversation, an ability to allow for silences and pauses, and a capacity for empathy will engage more successfully (Reissman, 1993; Wood & Kroger, 2015). Additionally, a thorough understanding of the cultural context, such as understanding the jargon, social cues, and other cultural properties, are critical to appropriately interpreting these conversations (Hutchby & Wooffitt, 2008; Johnstone, 2008; Potter & Wetherell, 2005).

Conversational and Discourse Analysis Formats

Data originating from conversational or discourse analysis evolve from naturally occurring conversations, ordinary and routine in many aspects (Sacks et al., 1974). These ordinary conversations may include talk among friends or colleagues, institutional talk in meetings or formalized settings, or classroom debates. The topic for any conversational or discourse approach is followed as the conversation emerges and flows or as a general topic that is introduced by the researcher and tracked to see where the participants take the topic. The tools for conversational discourse approaches range from loosely structured to completely unstructured; logbooks are used by the researcher to follow specific aspects of the conversation (predetermined before the topic is introduced). This tool also allows the researcher to record conversation threads and record observations, field notes, and other entries that relate to the verbal and nonverbal interactions in play. In this way, conversational and discourse approaches, stemming from the ethnographic process, are an adaptation and merging of unstructured interviews, observation, and ethnographic field note recordings. Additionally, video or audio recordings may be used as data sources for conversational and discourse data. Although most researchers prefer the video recordings in order to see the nonverbal interactions while they dissect the verbal interactions and patterns, it may be difficult to obtain permission for these types of recordings, and researchers may be limited to audio recordings as their primary data source (Van Dijk, 1997; Wiggins, 2017; Wood & Kroger, 2015).

Getting Started With a Basic Template

General Design Considerations

To maximize the conversational or discourse analysis process, a tool must be designed to accommodate all the types of data that will be recorded. The design process begins with a clear statement of the research purpose and questions, followed by a description of the research site, the participants and their profile characteristics (protecting their identities), and an initial typology for the major categories of inquiry. These categories should reflect the levels of discourse and interaction, the nonverbal behaviors and cues associated with all dialogue strands, and the inferences resulting from those associations. Not only should a tool capture what is apparent on the surface of the social and conversational interactions (profiles, setting, visible markings, and activities), but the researcher must also capture the underlying behaviors and meaning of the language. The goal is

Basic Conversational or Discourse Analysis Template

Template 5.1
Conversational or Discourse Analysis Log

Recording Log

TITLE OF STUDY: _____
DATE/TIME/DAY OF THE WEEK: _____
NUMBER OF PARTICIPANTS: _____ DESCRIPTION OF SETTING: _____

While observing conversation in process, the researcher will record features of the discourse related to the following categories:

Relates to:	Participants	Topics/content	Significant Statements	Beginning or ending of a conversation — catalyst or resolution
A. Individual behaviors, cues B. Group interactions and behaviors: Whom does this person relate to? Acknowledge? Defer to in a power distribution?				

(*Continued*)

(Continued)

Relates to:	Participants	Topics/content	Significant Statements	Beginning or ending of a conversation — catalyst or resolution
C. What is the pairing of speakers? Who tends to lead or follow in the conversation/discourse?				
D. Setting and use of space/objects as they relate to the conversation				
E. How does the conversation relate to the daily work, routines, and interactions of the speakers?				
F. How would you describe the organizational context for the interactions of participants? What are the challenges, issues that are in process during these conversations? How do participants refer to them?				
G. How do participants infer meaning to the topics they discuss?				
H. Demographic details				
I. Researcher reflections				

to study and convey how participants collaboratively coconstruct their talk via behaviors, events, internal tensions, external context, physical structures, and the layout of the environment in which participants are interacting. In the context of engaging in conversational and discourse analysis, the underlying premise is that talk occurs immediately after and in the course of a catalyst; something occurs to generate the conversation, and something follows the conversation to a conclusion that participants see as a resolution or as leverage for a new discussion. Their individual and collective viewpoints enrich their conversation. Therefore, the researcher's goal is to capture the overt and covert levels of discourse in order to understand the psychological and sociological perspectives inherent in their conversations (Fairlough, 2015; Gee, 2014b).

Template Variations and Challenges

There are very few instances where a template requires variation when collecting conversational or discourse data. If variations are required, the log that is used will naturally evolve from the basic template. Categories and sections added to or modified on a log will be specific to the study, the phenomenon explored, and the group of individuals under study.

Piloting Conversational and Discourse Analysis Tools

Piloting conversational and discourse tools requires a careful selection of a research site where participants resemble the final sample of participants but where those participants will not be included for live data collection. After obtaining permission to observe as part of a pilot test, researchers should spend adequate time listening to, watching, and analyzing ongoing conversations in order to test the categories on a log, making sure they provide adequate coverage. Once the pilot has been conducted, a review of the data will inform the researcher regarding any changes that are required.

Transforming Conversational and Discourse Data for Analysis

Recordings (video and audio) transformed into verbatim transcripts and audio/visual files comprise the raw data from these sources. Since CA and DA data are often supplementary to the primary data source(s), the analysis strategies are determined based on how the primary data are treated. The researcher has two choices when preparing CA and DA data for analysis; data may be analyzed within case or across case but will inevitably be converged with primary data for interpretation. Thematic, content, or heuristic analysis strategies are the most likely approaches for these data (Bischoping & Gazso, 2015; Gee, 2014b; Miles, Huberman, & Saldana, 2014; Rapley, 2011), and transcriptions of recordings are the starting point for analysis. As noted earlier, selecting reputable transcription services is important to secure quality products from which to complete data management, data reduction, and data coding and clustering.

The process of analyzing conversations or discourse data follows similar procedures with that of interview data or any data resulting from conversations among individuals (aside from focus groups, which are distinguished by their synergistic character). If conversations have been recorded, with the permission of the participants, the transcriptions that result provide the raw data for analysis. If permission was not granted by participants, the researcher's notes comprise the raw data for analysis. In both instances, the data must be reviewed and cleaned for verity and clarity before the next phase of analysis proceeds. Many researchers recommend strategies for conversational or discourse data as the best ways to capture the meaning of the group's language (Bischoping & Gazso, 2015; Fairlough, 2010; Gee, 2014a; ten Have, 2012; Heritage, 2004; Keller, 2013; Phillips & Hardy, 2015; Wooffitt, 2014). These approaches treat conversational and discourse data as narrative data, data which represent the context for the exchanges and reflections of the interactions among participants. In the process of organizing the various types of conversational data, the researcher's own typology will serve as the first step in organizing data for subsequent interpretation (Miles et al., 2014).

HIGHLIGHTS

Conversational and Discourse Analysis Tools	Conversational and discourse analyses comprise the study of conversations-in-process among participants at a site, where the nature of their dialogue is analyzed and interpreted.
	These types of data often serve as supporting or complementary data in a qualitative study.
	Qualitative researchers must be able to position themselves unobtrusively while listening to and observing participants in conversation and be able to record details and patterns in their speech.
	Formats for these data typically consist of conversations that occur naturally at a research site, and logs are used to record these exchanges.
	Logs are designed to capture behaviors, language, word patterns, interactions among participants, and notes about the setting/context for the conversations.
	Logs must be piloted in advance of live data collection to test the scope and breadth of the categories on the instrument.
	Data analysis for conversational and discourse data follows similar strategies to that of interview data analysis but focus on the meaning of the language and context for that language.

CHAPTER 6

Focus Group Moderator Guides

Focus groups are a research technique that collects data through group interaction on a topic determined by the researcher; in essence, it is the researcher-driven interest that provides the focus . . . but the data comes from the group interaction.

(Morgan, 1997, p. 6)

Focus Groups Defined

Focus group research is an increasingly popular qualitative data collection strategy and is used effectively by many educational researchers. One of the common pitfalls, however, is that many novice researchers (and even some experienced ones) may be unclear about what constitutes a focus group. There are several defining characteristics that distinguish a true focus group from other types of group interactions. As Krueger and Casey (2015) note, a focus group is "a carefully planned series of discussions designed to obtain perceptions on a defined area of interest in a permissive, non-threatening environment" (p. 2).

Focus group interviews, long accepted as a data collection strategy in social science research, originally surfaced in the 1940s as a method to test the public's response to World War II propaganda (Barbour, 2007). After years of use in business and marketing domains, focus groups have gained increasing acceptance and popularity in other research domains. As a qualitative research method, focus groups remain an ideal strategy for obtaining in-depth feedback regarding participants' attitudes, opinions, perceptions, motivations, and behaviors (Barbour & Kitzinger, 1999; Fern, 2001; Liamputtong, 2011; Morgan, 1997; Morgan & Krueger, 1998; Patton, 2015; Vaughn, Schumm, & Sinagub, 1996). As Morgan (1997) further notes, focus groups are useful when it comes to discovering not only what participants think but why they think as they do.

In general, focus groups comprise several distinct features. Small in size, they typically range from 6 to 12 participants; participants are purposefully selected, based on their commonalities, and often include participants originating from pre-existing groups. The discussion is focused, with a specific sequence of questioning, moving from general and broad open questions to specific and more complex or challenging questions. Sessions typically last between 60 and 90 minutes and are structured in their design, emphasizing consistent questioning across groups (Fern, 2001; Krueger & Casey, 2015; Vaugh et al., 1996).

As Morgan (1997) stresses, focus groups are collaborative interviews designed to capitalize on the group's evolving interaction. In this sense, the focus group differs from one-on-one interviewing in that each group generates its own outcomes and responses by virtue of being together. Yet, while the synergy of the group experience is ideal for cultivating rich and descriptive information about the topic under exploration, the process of designing and conducting these group interviews can also be challenging, especially for the beginning researcher. Even more critical, the researcher must develop an effective moderator's guide to facilitate the discussion and obtain meaningful, descriptive data.

Focus groups are not meant to consolidate individual interviews into a single, more efficient interview (Morgan, 1997; Morgan & Krueger, 1998). They differ from groups whose purpose is otherwise, such as therapy (patient-centered), presentations or debates (group-centered), or meetings/decision-making (leader-centered). The interactions of these specialized group discussions help participants further understand the topic of interest, yielding information not otherwise available through other data collection strategies. Unlike other types of groups, focus groups capitalize on the beliefs, ideas, and individual perceptions that surface as a result of a moderator-guided discussion framed in an interactive context.

There are numerous benefits as well as challenges associated with focus group research. The focus group's synergy generates responses among participants that build on the collective perspectives of group members; the give and take of the conversation brings issues to the surface resulting from these group's interactions. This process allows for extensive sharing, comparing, and elaboration among participants, offering the researcher an excellent and rich source of primary data. Conversely, focus groups may not provide an opportunity for sufficient depth of emotional responses and may yield only superficial results on a given topic. In this way, a researcher must be careful to determine if a topic is appropriate for focus group designs, especially if a topic is sensitive in nature. If a topic is particularly sensitive, it may not provide a safe environment for participants to fully or openly disclose their feelings on that subject (Krueger & Casey, 2015; Morgan & Krueger, 1998).

Additionally, some groups suffer from dominant or disruptive personalities who hijack the conversation, in which case the moderator must carefully manage and redirect the discussion. Finally, focus group results are not intended for generalizability but rather support the development of survey instruments that allow quantitatively derived results to be generalized (Barbour, 2007; Krueger & Casey, 2015; Liamputtong, 2011).

Focus Group Applications

Although often viewed as a self-contained exploratory, qualitative data collection strategy (Pizam, 1994), focus groups often supplement other data collection methods such as survey questionnaires, observations, and interviews (Morgan, 1997). Focus group interviews, therefore, can be integrated with qualitative projects in three different ways: (1) for use in exploratory/emergent designs, used when little is known about a topic or issue, and when focus groups can uncover the context, language, ideas, and expectations in more detail; (2) for use in self-contained designs, when focus group results can provide the sole source for data collection, viewed as a strategy to explore personal narratives, experiences, and shared experiences; and (3) for use as supplemental designs, when focus group results inform instrument design or serve as triangulation in mixed methods research designs. In this role, focus groups probe findings, corroborate similarities or differences, or reveal bias or inconsistencies in the preceding or subsequent findings (Liamputtong, 2011; Pizam, 1994). Additionally, focus groups may support the exploration and diagnosis of organizational dilemmas, employee satisfaction and workplace concerns, organizational planning and envisioning processes, program evaluation, and institutional needs-assessment (Krueger & Casey, 2015).

Requisite Skills and Characteristics of the Focus Group Moderator

All qualitative researchers must possess characteristics akin to those of a therapist, counselor, or coach; they need to listen, to communicate with sensitivity and compassion, and to elicit a participant's story in rich detail. The focus group moderator must demonstrate these skills but must also be able to step back from the conversation when appropriate.

Unlike the qualitative interviewer, who must consciously direct the conversation and the interaction between himself or herself and the interviewee, the focus group moderator must distinguish between starting the group's discussion and knowing when to turn that discussion over to the participants. As Flick (2009) suggests, the focus group moderator facilitates and guides rather than directs and controls the group's discussion. This skill is learned and developed over time and with practice; in addition, there are other essential moderator skills and characteristics that contribute to the focus group's success (Barbour, 2007; Krueger & Casey, 2015):

- Communicating the focus of the study's purpose and the way the findings will be used
- Focusing the discussion, keeping things on track, while still allowing the participants to direct the flow of the conversation
- Respecting all points of view
- Actively listening, effectively probing and clarifying participant comments
- Maintaining a nonjudgmental and nondefensive stance
- Actively encouraging everyone to speak and contribute to the discussion
- Managing difficult situations, difficult participants, and conflict within the group

Focus Group Types and Variations

A wide variety of focus group types are available to researchers. The single purpose session is the most common type, where a sole topic is explored with a single facilitator. Variations on single purpose focus groups, however, are numerous, as illustrated in Table 6.1. Depending on your research questions and your topic, one of these focus group types is ideally suited for your project (Kitzinger & Barbour, 1999; Morgan, 1997; Vaugh et al., 1996). Each of these focus group types uses a customized moderator's guide to reflect the purpose of the group session.

While this list is brief in its description of each type, many excellent resources are available to provide a full description of the varieties of focus groups, the scope of groups for different industries, the recommended

Table 6.1 Focus Group Types

Focus Group Type	Purpose
Single Purpose	Single topic, single moderator
Multiple Purpose	Single topic, multiple groups, multiple moderators; allows for across-group comparisons
Double-Layered Designs	Participants represent different strata of the same population, allows for specific focus on a subset or target group
Two-Way	Two different groups are paired, where one group actively discusses the topic at hand while the other group observes them and then discusses their interactions and conversation
Dual Moderators	Two moderators cofacilitate a single group where one moderator contributes procedural expertise and the other contributes content expertise
Dueling Moderators	Two moderators deliberately take opposing sides of a single issue to generate debate among participants
Brainstorming	Designed to generate preliminary or exploratory ideas for a project, plan, or event, using a process where participants combine roundtable comments with visuals or flipcharts to record ideas
Program Evaluation	Designed to assess and evaluate programs and recommend actions for improvement
Envisioning/Planning	Designed to envision or plan for a program's future goals, objectives, and actions, using a combination of flipcharts, visuals, lists, and other interactive tools
Online/Virtual/Teleconference	Participants interact and converse virtually, where the moderator runs the session from a platform that engages participants in conference calling, online meeting platforms, chat rooms, and other virtual meeting places, although these groups limit the moderator's ability to observe nonverbal cues in person

procedures for running focus group sessions and recruiting participants, and other considerations about these group discussions. The focus of this chapter is to assist researchers in the development of the moderator's guide to maximize the focus group session, which will support the study's objectives.

Getting Started With a Basic Template

General Design Considerations

The single purpose focus group moderator's guide remains the standard template for all focus group types. Minor modifications are all that are needed to transform the single purpose focus group template for use with other types such as the multiple purpose group, the double-layered design, and the online/teleconferencing group, all of which can adopt the basic moderator's guide template with ease. The same general design considerations apply to all of these focus group types, which depend on the development of comprehensible questions, a standardized sequence of those questions, probes that support the questions, and the time estimates allotted for each question.

Developing Focus Group Moderator Guide Questions

The purpose of the focus group is to cultivate synergy among group members and to get them to talk to each other, rather than to you as the moderator. Writing an initial list of ideas for your topics and possible questions is a good way to get started; review this list to ensure that your topics and questions directly address the purpose of your focus group and your research objectives. Sometimes a researcher may get sidetracked with a line of questioning that seems interesting but will not actually contribute to the session's focus or purpose.

As with other types of qualitative questions for other types of qualitative tools, it is important to maintain a neutral and straightforward tone with your questions. Avoid complicated or double-barreled questions; avoid value-laden or leading questions. Frame questions in a positive and nonjudgmental manner, and avoid bias or culturally insensitive language. Finally, keep the list of main content questions brief, no more than six, and ideally closer to four or five. These content questions comprise the heart of your discussion, but they are positioned between entry and exiting questions, so you need to make time for all the questions in your guide. Finally, maintain consistency in your questions across moderator's guides, should you require several different guides for several different group sessions where your participants or topics might deviate slightly. There may be instances where the mix of your groups will require slightly different variations on your questions, or your groups may represent slightly different views of a single topic. While variation may be necessary across your focus group guides, the key is to maintain consistency in at least 80% to 90% of your questions in order to compare and analyze the data uniformly (Billups, 2013).

Sequencing Focus Group Moderator Guide Questions

As most researchers will tell you, there is an established sequence for focus group questions. Krueger and Casey (2015) provide an excellent overview of this sequence and are supported by Barbour (2007), Fern (2001), Liamputtong (2011), and Morgan (1997). This sequence is aligned with time allotments for each question, which helps the moderator maintain a flow and a timeliness to the group's discussion. Earlier questions require less time and serve as entry into the discussion; later questions require more time and help participants ease into more complex or difficult topics.

The sequence of the typical focus group session includes the following progression:

- *Icebreaker/Opening Question* (60 seconds per person), where the questions are easy to answer, nonthreatening, and typically include simple introductory elements such as the person's first name, place of work/school, length of time at that location, or other basic pieces of information. The icebreaker also creates a climate where the moderator goes in a circle to make sure that each participant speaks out loud in front of the rest of the group; this action makes it easier for people to speak again during the remainder of the session. Going around the circle allows the moderator to establish a welcoming and inclusive feel to the group discussion by validating that everyone's voice is important to the discussion, regardless of how benign the icebreaker questions might seem.
- *Introductory Question* (60-90 seconds per person), where a topic is introduced that is related to the session topic but is still general enough to be easy to answer, nonthreatening, and encourages participants to contribute to one more round of speaking out loud and getting used to speaking in the group. This topic may cover a connection with the topic in some way, such as "tell us, in a minute to a minute

and a half, about a phrase or expression you would use to describe how you feel about teaching college students."

- *Transition Question* (1–2 minutes per person), where the discussion moves more specifically to the topic under study. In this phase, the general, broad conversation topics become more focused and more personal for participants. This is also the first time in the discussion when the moderator lets participants speak on their own impulses, rather than creating a structure for everyone to speak; it is the beginning of moving from a moderator-directed conversation to a participant-directed conversation, so the question must generate that inclination for participants. Questions are directly related to the session topic, and the moderator uses probes to encourage rich, detailed examples and descriptions of the participants' experiences.

- *Key or Content Questions* (open-ended, entire segment comprises approximately 40–45 minutes of the 90-minute focus group), where the real work of the focus group happens. The questions posed in this part of the discussion anchor the entire discussion; at least three or four substantial questions are asked by the moderator, and sometimes there might even be four to six questions. However, it is unlikely that more than four or five questions can be asked of participants without sacrificing the necessary details and stories that should come with participant perspectives. Asking too many questions may mean that insufficient details are provided, since participants may feel rushed by the moderator or may feel that there is not ample time for everyone to contribute to the discussion. Careful crafting of content questions is essential to answering your research questions and building a sense of synergy and ease in your group.

- *Debriefing/Concluding Question* (60-90 seconds or time determined by moderator depending on nature of the discussion), where the moderator determines that after the key questions have been covered, the group must exit the discussion safely and comfortably. There may be times when participants are deeply affected by a discussion or when their emotions or memories are disturbed in some way; creating a safe space within which participants can debrief, unload feelings, or process the discussion is another important element in focus group research. The questions created for this final phase should acknowledge the discussion that has just occurred but also return to a general level of discussion to depart from the intensity or intimacy of the main discussion.

Developing Probes

Each question in the focus group moderator's guide requires probes intended to keep the conversation on track. These probes assume different forms, such as asking for examples or stories; asking participants to rank or list things; or soliciting impressions, memories, ideas, goals, or aspirations. Regardless of the type of probe you use, it must relate to your questions; in other words, provide a time frame or time orientation (the future, the past, a specific time period) that helps the participants anchor their responses. Provide a specific reference to the question, and ask for further details for examples; probes such as "is there an example of that which you can share?" or "given what you just shared, how do you see that changing for you within the next year?" Other types of commonly used probes include those from the following list:

- What do you think about?
- Tell me more about that?
- How would you describe . . . ?
- What would you do if . . . ?
- Can you give me a specific example of that?
- Does anyone have something to add to what (*insert name*) said about that?

Probes are inserted in the guide after each key/content question and after the transition question; they are not typically used with icebreaker, introductory, or debriefing/concluding questions (Liamputtong, 2011; Stewart & Shamdasani, 2014).

Using the Moderator's Guide to Conduct the Session

The moderator's guide serves as a script for the moderator(s) and the note-taker/recorder. The guidelines for procedures, operations, and instructions play an important role in the focus group process. Therefore, when developing your guide, be sure to explicitly state your step-by-step procedures for welcoming participants, administering consent forms and presession questionnaires, distributing name tags and other materials, explaining the study's purpose and treatment and reporting of data, assurance of privacy or confidentiality of individuals and the findings, and the duration of the group session. These procedural guidelines assure participants that the session is well organized and adequate preparations have been completed. Before you commence the session, it is also important to establish group norms and ground rules. These activities take time, yet they are essential to providing a safe environment in which participants can speak freely and openly.

Thus, the focus group moderator's guide serves multiple purposes. First, the guide anchors the group discussion as the moderator uses the questions to direct the sequence and coverage of topics under study. Second, the guide serves as a procedural map, a blueprint of how the group discussion and process will unfold. Third, the guide provides a structure for operations, reminding the moderator and the note-taker/recorder or assistant moderator about the supporting activities that must occur to ensure a smooth process. Finally, the guide serves as the guidepost for the postsession debriefing, where the moderator and the note-taker/recorder or assistant moderator compare thoughts, notes, and initial impressions about the discussion. A carefully designed moderator's guide is, therefore, imperative to the success of the focus group (Billups, 2013).

Creating the Pre-Focus Group Profile Questionnaire

Most moderators administer a pre-focus group questionnaire to gather demographic data for each group. This presession survey is not only an excellent way to create a profile of the participants, but it also allows each person a "safe entry" into the focus group space and time

Template 6.1
Focus Group Moderator's Guide: Single Purpose

Information About the Focus Group

PARTICIPANTS (GENERAL): _____

MODERATOR: _____ GROUP: _____

DATE: _____ TIME: _____

PLACE: _____

Introduction, Process, Consent

- Introduce yourself.
- Review the study's purpose, how long you expect the focus group to take, and your plans for using the results.
- Note that the interview will be audio-recorded and that you will keep participants' identities confidential.
- Distribute any profile survey questionnaires at this time, as appropriate to your study.

Ground Rules

Ground rules and group norms are always established at the beginning of a focus group session to ensure mutual respect, consideration, and a supportive atmosphere for the discussion:

- All group members have a right to their viewpoints and opinions.
- All group members have a right to speak without being interrupted or disrespected by other group members.
- Group members will avoid dominating the conversation and will allow time for others to speak.

- The moderator has the right to guide the timing and flow of the session topics but will allow the group to determine the importance and focus of the conversation, as appropriate.
- Identities of group members will remain confidential; first names only will be used for name tags and in reference to one another during the session.

Questioning Sequence

1. Ice Breaker Question (60 seconds per participant)
2. Introductory Question (90 seconds per participant)
3. Transition Question (1–2 sentences in description per participant)
4. Content Questions
   a. Content #1
      i. Probes
   b. Content #2
      ii. Probes
   c. Content #3
      iii. Probes
   d. Content #4
      iv. Probes

Closing Question/Debriefing

5. What else would you like to tell me about?

Wrap Up and Thank You

- Thank you very much for your time today. I appreciated hearing your insights on this topic.

(If there is going to be a follow-up reflective process, please indicate that at this time.)

Note: An exemplar of the single purpose focus group moderator's guide is available in Appendix C.

Template 6.2
Focus Group Presession Participant Profile Questionnaire

Presession Questionnaire

Thank you for agreeing to participate in today's focus group session. Please take a moment to answer the following questions so we can better understand who you are, your work/industry background, and some of your preliminary thoughts about today's topic.

GENDER: _____

HIGHEST LEVEL OF EDUCATION: _____

PROFESSIONAL FIELD/INDUSTRY OF PRACTICE: _____

CURRENT POSITION (OPTIONAL): _____

Question

- Today, we will be talking about effective leadership practices; in your own words, please share your perspectives on what makes a leader effective.

_____

- In one sentence, please tell us how you would describe your *own* leadership style.

_____

Other Thoughts or Questions?

Is there anything else you wish to share with us regarding today's session? Please feel free to jot a few thoughts, questions, or observations in the space below:

Thank you! Please return this questionnaire to the moderator when you are finished.

to get a sense of the other attendees, the moderator(s), and the setting. In focus group research, every activity, every tool, and every connection has a purpose. A basic template for the presession participant profile questionnaire is listed below.

Focus Group Note-Taking Recording Sheet

Every focus group moderator uses a guide to direct the group conversation in a session. As part of the focus group team, every moderator works with a note-taker/recorder in order to document several levels of interactions, comments, and nonverbal behaviors of group members. These supplementary data provide a context for the focus group discussion and assist the moderator/note-taker team when debriefing at the end of the session. Just as a recording sheet is a necessary tool in the interviewing process, so it is with the focus group process.

There are three levels of documentation that the note-taker should watch for and record: group interactions, nonverbal behaviors on the

Template 6.3
Focus Group Note-Taking Recording Sheet

Interviewer notes/observations	Nonverbal cues	Quotable quotes

part of individuals and the group as a whole, and finally, the representative remarks made by group members that reflect the general tone and meaning of the focus group discussion. The note-taker recording sheet facilitates the documentation of these elements.

The note-taker must observe the way group members speak to one another, the tone of their voices, the speed of talking, and their body language responses to fellow group members and to the moderator. All of these cues indicate the various levels of comfort, honesty, and rapport that add to or detract from what they are saying. Using an instrument to record representative statements and group nonverbals is essential to understanding the group's discussion and should be used in conjunction with the moderator's guide.

Template Variations: Focus Group Moderator Guides by Focus Group Type

As noted above, there are several variations on the basic single purpose focus group moderator's guide. The variations listed in Table 6.2 are frequently used by moderators, each for a specific purpose and with specific features. A definition of each type, with a corresponding template, mirrors Table 6.1.

Table 6.2 Moderator Guides by Focus Group Variation/Type

Type	Key Feature	Description
Two-Way Design	2 guides	Two different groups are paired, where one group actively discusses the topic at hand while the other group observes them and then discusses their interactions and conversation
Dual Moderators	1 guide, 2 roles	Two moderators cofacilitate a single group where one moderator contributes procedural expertise and the other contributes content expertise

Dueling Moderators	1 guide, 2 roles	Two moderators deliberately take opposing sides of a single issue to generate debate among participants
Brainstorming/ Envisioning	1 guide, w/ visuals	Designed to generate preliminary or exploratory ideas for a project, plan, or event, using a process where participants combine roundtable comments with visuals or flipcharts to record ideas; envisioning a plan for the future
Program Evaluation	1 guide, w/ visuals	Designed to assess and evaluate programs and recommend actions for improvement

Template 6.4.1
Focus Group Moderator's Guide for Two-Way Designs (Group #1)

In the two-way focus group, one group actively participates in the discussion, moderated by the facilitator; the other group observes the discussion, after which the moderator solicits their observations and perceptions about the first group's interactions. Therefore, there are two templates used for the two-way design: one template is designated for Group #1, and the other is designated for Group #2.

Two-Way Design Template: Group #1

Information About the Focus Group

PARTICIPANTS (GENERAL): _____

MODERATOR: _____ GROUP: _____

DATE: _____ TIME: _____

PLACE: _____

(Continued)

(Continued)

Introduction, Process, Consent

- Introduce yourself.
- Review the study's purpose, how long you expect the focus group to take, and your plans for using the results. Explain the purpose of the two-way design and the phases of the process (Group #1 discussion, observed by and then reviewed by Group #2).
- Note that the interview will be audio-recorded and that you will keep their identities confidential.
- Distribute any profile survey questionnaires at this time, as appropriate to your study.

Ground Rules

Ground rules and group norms are always established at the beginning of a focus group session to ensure mutual respect, consideration, and a supportive atmosphere for the discussion:

- All group members have a right to their viewpoints and opinions.
- All group members have a right to speak without being interrupted or disrespected by other group members.
- Group members will avoid dominating the conversation and will allow time for others to speak.
- The moderator has the right to guide the timing and flow of the session topics but will allow the group to determine the importance and focus of the conversation, as appropriate.
- Identities of group members will remain confidential; first names only will be used for name tags and in reference to one another during the session.

Questioning Sequence

1. Ice Breaker Question (60 seconds per participant)
2. Introductory Question (90 seconds per participant)
3. Transition Question (1–2 sentences in description per participant)
4. Content Questions

a. Content #1
      i. Probes
   b. Content #2
      ii. Probes
   c. Content #3
      iii. Probes
   d. Content #4
      iv. Probes

Closing Question/Debriefing

5. What else would you like to tell me about? _____

Wrap Up and Thank You

- Thank you very much for your time today. I appreciated hearing your insights on this topic.
  (If there is going to be a follow-up reflective process, please indicate that at this time to prepare group members.)

Template 6.4.2
Focus Group Moderator's Guide for Two-Way Designs (Group #2)

Two-Way Design Template: Group #2

Information About the Focus Group

PARTICIPANTS (GENERAL): _____

MODERATOR: _____ GROUP: _____

DATE: _____ TIME: _____

PLACE: _____

(Continued)

(Continued)

Introduction, Process, Consent

- Introduce yourself.
- Review the study's purpose, how long you expect the focus group to take, and your plans for using the results. Explain the purpose of the two-way design and the phases of the process (Group #1 discussion, observed by and then reviewed by Group #2).
- Note that the interview will be audio-recorded and that you will keep their identities confidential.
- Distribute any profile survey questionnaires at this time, as appropriate to your study.

Ground Rules

Ground rules and group norms are always established at the beginning of a focus group session to ensure mutual respect, consideration, and a supportive atmosphere for the discussion:

- All group members have a right to their viewpoints and opinions.
- All group members have a right to speak without being interrupted or disrespected by other group members.
- Group members will avoid dominating the conversation and will allow time for others to speak.
- The moderator has the right to guide the timing and flow of the session topics but will allow the group to determine the importance and focus of the conversation, as appropriate.
- Identities of group members will remain confidential; first names only will be used for name tags and in reference to one another during the session.

Questioning Sequence

1. Ice Breaker Question (60 seconds per participant) (This question should focus on their initial impressions of Group #1's interactions, so asking about a word or phrase that members would use to describe that session would be an effective icebreaker.)

2. Introductory Question (90 seconds per participant) (This question should follow up on the icebreaker by asking members to describe the ways in which they felt that Group #1 interacted or communicated.)
3. Transition Question (1–2 sentences in description per participant) (This transition question should focus on the essence of the discussion in Group #1 and the extent to which Group #2 members feel that they addressed the questions, shared experiences, and/or effectively interacted and communicated.)
4. Content Questions (The moderator uses the content questions to explore the various observations of Group #2 regarding the interactions, communication style, nonverbal behaviors, and depth of content evidenced in Group #1.)
    a. Content #1
        i. Probes
    b. Content #2
        ii. Probes
    c. Content #3
        iii. Probes
    d. Content #4
        iv. Probes

Closing Question/Debriefing

5. What else would you like to tell me about?

Wrap Up and Thank You

- Thank you very much for your time today. I appreciated hearing your insights on this topic.

(If there is going to be a follow-up reflective process, please indicate that at this time to prepare group members.)

Template 6.5
Focus Group Moderator's Guide for Dual Moderators

In the dual moderator focus group, two moderators work in tandem to facilitate the group's discussion and interactions. Typically, the moderator with content expertise asks the questions while the moderator with procedural expertise ensures that all the questions are asked and helps to keep the discussion on track. One moderator's guide is used for this type of session, but the moderators agree on assigning roles for the session. For instance, the content moderator (CM) may ask all the main questions, but the procedural moderator (PM) may probe the responses. The PM may also oversee the completion of the profile questionnaire and consent forms while the CM welcomes members to the session.

Information About the Focus Group

PARTICIPANTS (GENERAL): _____

MODERATOR: _____ GROUP: _____

DATE: _____ TIME: _____

PLACE: _____

Introduction, Process, Consent

- Introduce yourself.
- Review the study's purpose, how long you expect the focus group to take, and your plans for using the results.
- Note that the interview will be audio-recorded and that you will keep their identities confidential.
- Distribute any profile survey questionnaires at this time, as appropriate to your study.

Ground Rules

Ground rules and group norms are always established at the beginning of a focus group session to ensure mutual respect, consideration, and a supportive atmosphere for the discussion:

- All group members have a right to their viewpoints and opinions.
- All group members have a right to speak without being interrupted or disrespected by other group members.
- Group members will avoid dominating the conversation and will allow time for others to speak.
- The moderator has the right to guide the timing and flow of the session topics but will allow the group to determine the importance and focus of the conversation, as appropriate.
- Identities of group members will remain confidential; first names only will be used for name tags and in reference to one another during the session.

Questioning Sequence

1. Ice Breaker Question (60 seconds per participant) (CM)
2. Introductory Question (90 seconds per participant) (CM)
3. Transition Question (1–2 sentences in description per participant) (PM)
4. Content Questions
   a. Content #1 (CM)
      i. Probes (PM)
   b. Content #2 (CM)
      ii. Probes (PM)
   c. Content #3 (CM)
      iii. Probes (PM)
   d. Content #4 (CM)
      iv. Probes (PM)

Closing Question/Debriefing

5. What else would you like to tell me about?

_____ (CM)

Wrap Up and Thank You

- Thank you very much for your time today. I appreciated hearing your insights on this topic.

(If there is going to be a follow-up reflective process, please indicate that at this time to prepare group members.)

Template 6.6
Focus Group Moderator's Guide for Dueling Moderators

In the dueling moderator focus group, two moderators deliberately take opposing views on a topic in order to engage the group and generate conversation about these opposing viewpoints. The same moderator's guide can be used, with the content questions assigned to Moderator #1 (M1) and Moderator #2 (M2) to juxtapose their stances.

Information About the Focus Group

PARTICIPANTS (GENERAL): _____

MODERATOR: _____ GROUP: _____

DATE: _____ TIME: _____

PLACE: _____

Introduction, Process, Consent

- Introduce yourself.
- Review the study's purpose, how long you expect the focus group to take, and your plans for using the results.
- Note that the interview will be audio-recorded and that you will keep their identities confidential.
- Distribute any profile survey questionnaires at this time, as appropriate to your study.

Ground Rules

Ground rules and group norms are always established at the beginning of a focus group session to ensure mutual respect, consideration, and a supportive atmosphere for the discussion:

- All group members have a right to their viewpoints and opinions.
- All group members have a right to speak without being interrupted or disrespected by other group members.

- Group members will avoid dominating the conversation and will allow time for others to speak.
- The moderator has the right to guide the timing and flow of the session topics but will allow the group to determine the importance and focus of the conversation, as appropriate.
- Identities of group members will remain confidential; first names only will be used for name tags and in reference to one another during the session.

Questioning Sequence

1. Ice Breaker Question (60 seconds per participant)
2. Introductory Question (90 seconds per participant)
3. Transition Question (1–2 sentences in description per participant)
4. Content Questions
    a. Content #1 (Moderator #1)
        i. Probes
    b. Content #2 (Moderator #2)
        ii. Probes
    c. Content #3 (Moderator #1)
        iii. Probes
    d. Content #4 (Moderator #2)
        iv. Probes

Closing Question/Debriefing

5. Synthesis: Reviewing the different viewpoints and the final assessment of the group regarding the discussion (Moderators #1 and #2 cofacilitate this last question)
6. What else would you like to tell us about today's discussion? _____

Wrap Up and Thank You

- Thank you very much for your time today. I appreciated hearing your insights on this topic.

(If there is going to be a follow-up reflective process, please indicate that at this time to prepare group members.)

Template 6.7
Focus Group Moderator's Guide for Brainstorming/Envisioning

Brainstorming focus groups are fluid, open discussions that involve considerable interaction and physical movement, using newsprint, Post-It notes, and other interactive tools to create an environment of creative idea-generating discussion. These groups require fewer questions but greater use of lists and visual aids to advance the discussion.

Similar to the brainstorming focus group is the envisioning/planning group, where the emphasis is on soliciting member views about the creation of an organizational plan or vision. The moderator acts as a planning facilitator in many ways; the group is asked very specific questions regarding organizational mission, vision, values, goals, and action steps. These sessions often include many of the same visual aids as you might see in a brainstorming group, such as newsprint, whiteboard, lists and rankings, and Post-It notes. These various tools help participants articulate and share their ideas about future directions and aspirations for the organization or unit they are discussing. The templates are interchangeable in many ways and are presented below as a single template design.

Information About the Focus Group

PARTICIPANTS (GENERAL): _____

MODERATOR: _____ GROUP: _____

DATE: _____ TIME: _____

PLACE: _____

Introduction, Process, Consent

- Introduce yourself.
- Review the study's purpose, how long you expect the focus group to take, and your plans for using the results.
- Note that the interview will be audio-recorded and that you will keep their identities confidential.
- Distribute any profile survey questionnaires at this time, as appropriate to your study.

Ground Rules

Ground rules and group norms are always established at the beginning of a focus group session to ensure mutual respect, consideration, and a supportive atmosphere for the discussion:

- All group members have a right to their viewpoints and opinions.
- All group members have a right to speak without being interrupted or disrespected by other group members.
- Group members will avoid dominating the conversation and will allow time for others to speak.
- The moderator has the right to guide the timing and flow of the session topics but will allow the group to determine the importance and focus of the conversation, as appropriate.
- Identities of group members will remain confidential; first names only will be used for name tags and in reference to one another during the session.

Questioning Sequence

1. Ice Breaker Question (60 seconds per participant)
2. Introductory Question (90 seconds per participant)
3. Transition Question (1–2 sentences in description per participant) (Visual aids may be used at this point.)
4. Content Questions (These questions often involve group interaction around visual tools to articulate thoughts to help group members not only hear what others are saying but to see how those viewpoints can be organized for mission/vision updates, goals, and action plans.)
5. Content Questions (Visual aids are commonly used for all content questions.)
    a. Content #1
        i. Probes
    b. Content #2
        ii. Probes
    c. Content #3
        iii. Probes
    d. Content #4
        iv. Probes

(Continued)

(Continued)

Closing Question/Debriefing

6. What else would you like to tell me about?
   _____

Wrap Up and Thank You

- Thank you very much for your time today. I appreciated hearing your insights on this topic.

(If there is going to be a follow-up reflective process, please indicate that at this time to prepare group members.)

Template 6.8
Focus Group Moderator's Guide for Program Evaluation

In the program evaluation focus group, the moderator solicits participant views of a program and works toward recommendations for action and improvement. Visual aids are often used with these groups, such as newsprint, whiteboards, charts, lists, rankings, and other tools that help participants articulate their impressions, assessments, and recommendations for future action. In many instances, participants are allowed to preview program information prior to the session to help them prepare for the questions and the activities.

Information About the Focus Group

PARTICIPANTS (GENERAL): _____

MODERATOR: _____ GROUP: _____

DATE: _____ TIME: _____

PLACE: _____

Introduction, Process, Consent

- Introduce yourself.
- Review the study's purpose, how long you expect the focus group to take, and your plans for using the results.

- Note that the interview will be audio-recorded and that you will keep their identities confidential.
- Distribute any profile survey questionnaires at this time, as appropriate to your study.

Ground Rules

Ground rules and group norms are always established at the beginning of a focus group session to ensure mutual respect, consideration, and a supportive atmosphere for the discussion:

- All group members have a right to their viewpoints and opinions.
- All group members have a right to speak without being interrupted or disrespected by other group members.
- Group members will avoid dominating the conversation and will allow time for others to speak.
- The moderator has the right to guide the timing and flow of the session topics but will allow the group to determine the importance and focus of the conversation, as appropriate.
- Identities of group members will remain confidential; first names only will be used for name tags and in reference to one another during the session.

Questioning Sequence

1. Ice Breaker Question (60 seconds per participant)
2. Introductory Question (90 seconds per participant)
3. Transition Question (1–2 sentences in description per participant)
4. Content Questions (These questions often involve group interaction around visual tools to articulate thoughts to help group members not only hear what others are saying but to see how those viewpoints can be listed, ranked, categorized, or clustered for future action.)
   a. Content #1
      i. Probes
   b. Content #2
      ii. Probes

(Continued)

> (Continued)
>
>   c. Content #3
>   iii. Probes
>   d. Content #4
>   iv. Probes
>
> ### Closing Question/Debriefing
>
> 5. What else would you like to tell me about?
>
> ---
>
> ### Wrap Up and Thank You
>
> - Thank you very much for your time today. I appreciated hearing your insights on this topic.
>
> (If there is going to be a follow-up reflective process, please indicate that at this time to prepare group members.)

Dyadic Interviews: The Facilitated 2-Participant Interview

On the continuum of the individual interview to the synergistic group interview (focus group), lies the dyadic interview. As Morgan (2016) notes:

> Traditionally, qualitative interviews have involved a single participant in one-to-one interviews or several participants in a focus group. There is thus an interesting gap in the size range, which does not include interviews that involve pairs of participants. Dyadic interviews fill that gap. (p. 9)

Dyads are researcher-facilitated two-person interviews, where the researcher serves as a moderator to encourage dialogue between the participants and to generate discussion on predetermined topics (Morgan, 2016; Morgan, Ataie, Carder, & Hoffman, 2013). Dyads are also known as joint, peer, paired, and two-person interviews. Ultimately, the dyadic conversation allows for a closer and deeper connection between participants due to the smaller number of persons involved; the conversation is more actively coconstructed between participants. Many of the characteristics of conversational analysis interactions find a place in the

Template 6.9
Moderator's Guide for Dyadic Interviews

Dyads are researcher-facilitated interviews with two participants, where the researcher serves as a moderator to encourage synergistic dialogue between the two participants and to generate discussion on predetermined topics (Morgan, 2016). The template follows the same structure as the one designated for a single purpose focus group.

Information About the Focus Group

PARTICIPANTS (GENERAL): _____

MODERATOR: _____ GROUP: _____

DATE: _____ TIME: _____

PLACE: _____

Introduction, Process, Consent

- Introduce yourself.
- Review the study's purpose, how long you expect the interview to take, and your plans for using the results.
- Note that the interview will be audio-recorded and that you will keep their identities confidential.
- Distribute any profile survey questionnaires at this time, as appropriate to your study.

Ground Rules

Ground rules and group norms are always established at the beginning of an interview session to ensure mutual respect, consideration, and a supportive atmosphere for the discussion:

- All group members have a right to their viewpoints and opinions.
- All group members have a right to speak without being interrupted or disrespected by other group members.

(Continued)

(Continued)

- Group members will avoid dominating the conversation and will allow time for others to speak.
- The interviewers have the right to guide the timing and flow of the session topics but will allow the group to determine the importance and focus of the conversation, as appropriate.
- Identities of group members will remain confidential; first names only will be used for name tags and in reference to one another during the session.

Questioning Sequence

1. Ice Breaker Question (1–2 sentences per participant)
2. Introductory Question (general topic)
3. Transition Question (general topic)
4. Content Questions (Questions become more specific progressively through the content section.)
    a. Content #1
        v. Probes
    b. Content #2
        vi. Probes
    c. Content #3
        vii. Probes
    d. Content #4
        viii. Probes

Closing Question/Debriefing

5. What else would you like to tell us about?

_____

Wrap Up and Thank You

- Thank you very much for your time today. We appreciated hearing your insights on this topic.

(If there is going to be a follow-up reflective process, please indicate that at this time to prepare group members.)

dyadic interview; the elements of "turn-taking," response and reaction, and pairing of ideas is evident in a dyad.

Like the focus group, the interactions of all participants form the basis for the interview. These interactions develop and expand based on participant perceptions of the topic and the questions posed by the facilitator (Morgan, 2016). The synergy of these two-person conversations means that dyadic interviews share some properties with focus group discussions while also reflecting some of the properties of in-person, 1-on-1 interviews (there are also triadic interview formats, but they are less known and little used; Morgan, 2016, p. 15). The researcher uses a special guide to conduct this type of interview, but the overlap with the focus group moderator's guide is important to consider. While similar to the moderator's guide in its structure, designed to capture the conversation flow, the technique of funneling is embedded in the guide (Morgan, 2016, p. 63). In this approach, the facilitator ensures that the topics first covered in the discussion are general, nonthreatening, and encourage open dialogue between participants; as familiarity ensues, the funnel closes, and the questions become more specific. Because there are fewer participants in the discussion, each individual has more time to speak and more time to process as the conversation proceeds. The dyadic guide, therefore, eliminates some of the time restrictions inherent in the focus group guide, and instead allows for longer topic coverage and deeper discussion.

Piloting Focus Group Moderator Guides

As with all instruments and tools, moderator guides should be piloted prior to live data collection. A pretest should be conducted with a small group of three or four individuals who resemble but will not be included in the final participant group. If there are multiple guides used in a particular design, all the guides should be pretested.

The pilot process helps the researcher understand whether the focus group questions are clear, in the right order, are redundant or overlapping in any way and to what extent they adhere to the appropriate time allotments. The pretest also informs the researcher about the instructions, procedures before and after the session, and the partnership with the note-taker/recorder, or assistant moderator. Once the pilot test is complete, the moderator's guide is revised and ready for final implementation.

Transforming Focus Group Data for Analysis

As a data source, focus groups generate rich, descriptive information to provide participant perspectives on a topic, but this data is extensive and often overlapping. When faced with the volume of focus group data, many beginning qualitative researchers assume that analyzing it is similar to analyzing other types of qualitative data. This assumption, however, dilutes the distinctions and richness of information that results from the focus group discussion (Billups, 2016).

In order to prepare your data for analysis, you must be aware of how the data will be treated during analysis. Therefore, transforming your focus group data for analysis includes the following:

1. After a review of your research purpose and research questions, organize all transcripts, recorder notes, and debriefing/member checking notes from all the focus group sessions, and read through them once without stopping to make notes or codes (if you conducted multiple focus groups, this will take some time).

2. After a period of at least 24 hours (to allow for processing and to avoid "recall confusion"), review all materials a second time, and create margin notes about the distinctions between individual comments and group interactions/dynamics.

3. Read all materials a third time, and begin to make a preliminary code list, starting with broad categories; continue this process until all chunks of data have been assigned preliminary codes.

4. The next level of analysis seeks to find repeated patterns of meaning at the group interaction level, by reviewing the data and juxtaposing the contrasts between individual perspectives and group interactions (Are there key differences between individual views and the actual group perspective? Is one individual forcefully asserting their viewpoint at the expense of the general mood or sentiments of the group as a whole?).

Next, using Krueger and Casey's guidelines (2015, p. 147), consider how your data may be analyzed using the following six aspects:

- Frequency—how often was something mentioned, and what is the relevance of and weight of all statements?
- Specificity—seek detailed comments and identify comparisons across cases and within cases.
- Emotion/nonverbal—record nonverbal and emotional responses, and corroborate them with participant comments (some participants become emotional when they feel they are not being heard, while at other times emotional commentary serves as a catalyst for group consensus).
- Extensiveness—consider how many participants said similar things versus how many times a single person reiterated an idea (look across your groups for variations on extent).
- Outliers—review all outlier statements to determine whether they reflected something worth pursuing (this is essential as every discussion will include some comment that seems off topic but may have value).
- Relationships—what were the relationships between statements and content, and what did they mean? Were alliances formed among group members, and how did this affect the conversation?

Classic Approach

While many qualitative researchers debate the best strategies for analyzing focus group data, Krueger and Casey (2015) are the acknowledged experts in this realm. Their Classic Approach strategy is one of the most common and manageable processes that can be applied to focus group data (Krueger & Casey, 2015). The nature of focus group analysis is reflected in its systematic and continual nature; analysis begins with the first completed session and is ongoing with all subsequent sessions. One of the process's main challenges is to balance the voices of the individuals with the collective and emergent perspectives of the group, a process that is fluid and changeable as the discussion evolves. Reflecting on the interactions among the participants allows for the complexities of group dynamics to inform data analysis. Unlike individual interview data or observation/document analysis data, issues of frequency, specificity, emotion, and extent are key elements. Additionally, deviant or outlier case analyses play an important role in understanding these data.

Analytical Frameworks

Using the Classic Approach analytic strategy provides for an overlay of a choice of several frameworks, allowing for deeper interpretation of the data. The choice of a framework is dependent on the nature and purpose of your focus group research objectives and allows you to examine your data from different perspectives, allowing for alternative explanations. As a framework for understanding how to prepare your focus group data for analysis, three of the most frequently applied frameworks are outlined in the following (Krueger & Casey, 2015, pp. 157–158).

Key Concepts Framework. The goal of this framework is to identify the factors of central importance, common to most of the participants in the discussion, which aids in understanding how participants view the topic in question. This framework focuses on the centrality of comments rather than on the extremes; the most commonly stated or agreed upon concepts are important in identifying the core elements. This framework is commonly applied when designing or assessing new or existing programs or when seeking to address customer/client needs; outliers or extreme perspectives are minimized in this framework.

Critical Incidents Framework. The objective here is to identify critical events that have shaped participant decisions or actions, grounded by the emotional or organizational forces surrounding those incidents. There is less emphasis on patterns in the data (themes) and more emphasis on extracting the details of the incidents to explain what has happened in the organization or group. This approach is often used when exploring program successes or failures, organizational functioning or dysfunction, or identifying triggering causes or catalysts in a group process.

Testing Alternatives Framework. The primary goal for this framework is to identify the most preferred choice among several alternatives and to avoid succumbing to the most assertive voice in the group; seeking and identifying group consensus is essential in this framework. The most common application of this framework is intended for pilot testing programs or services, curriculum, or academic processes. Participants are provided with a set of options to consider, and the group facilitator manages the conversation to elicit opinions and eventual group consensus. The brainstorming or scenario exploration exercise acts as a prelude to organizational planning or goal setting, and this framework aids in clarifying important organizational objectives.

HIGHLIGHTS

Focus Group Moderator Guides

Focus groups are carefully planned and guided synergistic group interviews.

Focus groups are used to support exploratory, self-contained, triangulation, or supplementary data collection.

Focus group moderators must be adept at group management, guiding discussions without controlling those discussions and respectfully letting the group direct the flow of topics.

Types of focus groups include single and multiple purpose groups, double-layered design groups, two-way designs, dual moderators, dueling moderators, brainstorming groups, program evaluation groups, envisioning/planning groups, and online/virtual/teleconference groups.

Single and double purpose groups, double-layered design groups, and online/virtual/teleconference groups use the same basic moderator's guide; other variations use the basic template as a starting point to customize guides for specific group types.

Moderator guides follow a standard question sequence with icebreaker, introductory, transition, content questions, and closing/debriefing questions.

Dyadic interviews are a form of a synergistic discussion and use a moderator's guide similar to that of other focus group types; triad interviews are also an option.

Piloting the focus group moderator's guide requires the selection of approximately three individuals to pretest the guide.

Focus group data analysis requires special strategies specific to the approach, typically applying the Classic Approach and the overlay of three distinct frameworks.

CHAPTER 7

Observation Tools

Qualitative observation refers to data that are observed with our senses: sight, smell, touch, taste, hearing. These observations do not involve measurements or numbers. Instead, they comprise behaviors, non-verbal actions and interactions, shapes and textures of objects, the physical environment and setting, and anything else that may contextualize the study of individuals in their natural setting.

(Author)

The observer is part spy, part voyeur, part fan, part member.

(Van Maanen, 1978, p. 346)

Observation Defined

Lincoln and Denzin (2008) capture the essence of observation: "Going into a social situation and looking is another important way of gathering materials about that social world" (p. 48). Marshall and Rossman (1989) extend this definition by adding that observation is "the systematic description of events, behaviors, and artifacts in the social setting chosen for study" (p. 79). What does it mean to observe others? What is the distinction between looking and seeing? Some researchers suggest that the value of observation lies in the residual effects of what a place or culture or institution leaves behind, or what it records for others to see or study (Angrosino, 2014; Lofland & Lofland, 1995).

Most researchers agree that observation is a systematic data collection approach and that researchers must use all of their senses to discern people in their natural settings or in naturally occurring situations (Angrosino, 2014; Cohen & Crabtree, 2006; DeWalt & DeWalt, 2011; Lofland & Lofland, 1995). Through these systematic observations of the events, behaviors, and artifacts in the social setting, the researcher learns about the meanings attached to those things. Therefore, an assumption is made that behavior is purposive and expressive of the deeper values in the context for the individuals being observed. As a form of field work, the researcher must engage in "active looking, improving memory, informal interviewing, writing detailed field notes" (DeWalt & DeWalt, 2002, p. viii). In all of this work, the researcher must blend into the environment in such a way as to remain unobtrusive but intimately observant and aware of what is going on (DeWalt & DeWalt, 2011).

In this range of participantness (Patton, 2002, 2015), the researcher's involvement is represented on a continuum from full immersion to detached observer. Gold (1958) first provided a description of this continuum, noting that there are four stances that the observer can adopt: (1) complete participant, someone who is a member of the very group under study, (2) observer as quasiparticipant, someone who cares more for observing than participating as a group member, (3) researcher as participant but not a true member of the group, and (4) complete observer, someone with no membership in the group under study. At the one end, with full participation, the researcher achieves full disclosure with the observed participants; at the other end, there is complete secrecy and a level of deceit involved in observing those who do not know they are under study. The researcher must determine to what extent he or she will become involved (Spradley, 2016; Kawulich, 2005). In all cases, the observer must note not just what is happening but what is *not* happening, as well; the overt and the covert frame the observer's work.

Observation Applications

Observation data are equally viable as a primary or secondary/supplementary data source. Observation is normally associated with ethnographies, associated with the use of field notes to record details of the research setting, participant behaviors, and other contextual information. The practice can, however, support other research designs, such as descriptive/interpretive, case study, and grounded theory studies (Bryant & Charmaz, 2007). There are a variety of occasions where observation data are valuable:

- For relatively unexplored topics
- Where little is already known about the behavior of people in a particular setting or context

- When the interpretation of a setting is critical to understanding the phenomenon under study

- When the researcher explores what people say about their behaviors versus how they actually behave

- When 1-on-1 or group interactions comprise an important element in data triangulation for a study

- When researchers may be able to observe things about their participants that those participants may be unwilling to share in interviews or group discussions (Angrosino, 2014; Lofland & Lofland, 1995)

Observation data add value to a study as a supplementary source of data and can enrich a study in many ways. As part of a triangulation effort, observation can also add a different perspective when compared with the spoken or written findings.

Requisite Skills and Characteristics of the Observer

The value of observation is the intentionality of the process; the researcher should carefully plan for and prepare tools to collect data in a systematic manner. Becoming familiar with the setting beforehand is essential; listening to conversations and observing interactions before beginning the note-taking process will allow the researcher to absorb what is "normal" in that environment. This delay ensures that the researcher does not jump to conclusions about what he or she is observing.

Angrosino and de Perez (2000) and Bernard (1994) note that one of the basic skills an observer must employ is that of patience and the ability to be unobtrusive, open, nonjudgmental, and interested in learning more about the participants under study. The observer must be adept at careful, detailed note-taking, looking for patterns in language, behaviors, interactions, and where to draw the boundaries around what should be observed and what should be excluded. Merriam (2002) stresses the importance of shifting from a broad perspective to an increasingly focused perspective and going back and forth to understand the relationship between the context and the meaning of the interactions within that context. Several researchers emphasize that the best results occur when the researcher considers those he or she is observing as collaborators in the endeavor (DeWalt & DeWalt, 2011; Eppley, 2006; Kawulich, 2005).

If there are obstacles of language or culture, researchers must make every effort to acculturate themselves with those aspects of the situation, to increase the verity of the observations, and to honor the characteristics of the field. The stance of the observer is crucial to the emergence of the setting, scene, and activities of the participants, particularly as the observations are recorded from the field where the immediacy of the setting prevails. Researcher-observers must also be skilled at the practice of reflexivity, since observation requires what is seen to be filtered through the researcher's lens, viewpoint, and inherent biases (Lofland & Lofland, 1995).

Observation Formats

In the context of Gold's (1958) continuum, there are essentially two accepted types of observation—participant and nonparticipant observation—that dictate the observation format a researcher will employ (DeWalt & DeWalt, 2011). Although participant observation is more easily defined than nonparticipant, both types of observation require prolonged engagement in a setting, typically where participants are in a social situation or formalized environment. The researcher observes how participants express themselves, interact with one another, respond to social stimuli, make sense of their environment, and govern their own behaviors, language, and physical adaptation to the setting (Angrosino, 2014).

Participant observation demands firsthand involvement in the social world under study; it allows a researcher to hear, see, and experience reality from the participant's perspective (Spradley, 2016). Fetterman (1998) notes that participant observation "combines participation in the lives of the people being studied with the maintenance of a professional distance that allows adequate observation and recording of data" (pp. 34–35). This understanding is only possible with prolonged engagement, an intentional strategy to spend considerable time in the setting and learn about the daily life and routines of the group being observed.

Conversely, nonparticipant observation occurs when the researcher is directly excluded from or maintains limited interaction with the activities, behaviors, and language of the group under study and observes as a third-party (Eppley, 2006). This distance affects the researcher's ability to understand some levels of the interactions between group members if he or she does not have prior knowledge of the group. On the other hand, if the researcher observes a group for which he or she has extensive knowledge, observation is less restrictive. Regardless, this difference in observation status does not change the need for prolonged engagement with the group; the amount of time spent in the setting on a regular basis and the duration of the study is just as important to the nonparticipant observer as it is to the fully immersed participant observer. The important distinction, and one that affects the nature of interpreting the findings, is whether relationships are developed with participants and to what extent trust can be established with them.

While the observation rubrics may remain the same, nonparticipant observation requires supplementary tools to capture observation data. Due to the distance between researcher and participant, audio and video recording devices may play a role that the participant observer might otherwise play. The challenge of the nonparticipant observer who depends on audio recordings alone suggests that the wealth of non-verbal behaviors and cues are lost to possible interpretation; capturing the details of nuanced behaviors and body language is important to the deeper understanding of the participants and their setting. Additionally, if people are to be observed in a closed setting, such as a classroom or meeting space, and the researcher has requested audio or video recordings of the proceedings, the distinction clearly denotes the researcher as a nonparticipant observer. Aside from that distinction, there is some debate about how clearly the line is drawn between participant and non-participant observation in qualitative studies (Cohen & Crabtree, 2006; Spradley, 2016).

Getting Started With the Basic Template

General Design Considerations

Observational rubrics must facilitate the recording of information from several different perspectives to secure a holistic picture of the phenomenon. Angrosino and de Perez (2000) and Werner and Schoeppfle (1987) highlight three aspects of observation that must be incorporated into the observation rubric: (1) descriptive observation, (2) focused observation, and (3) selective observation.

Merriam (2002) adds to this list by recommending a device that captures the following aspects of the observed reality: (1) description of the physical environment (drawing a map, a visual representation of the setting), supported by written descriptions of those environments; (2) detailed descriptions of the participants; (3) identification of the frequencies and durations of interactions among participants; (4) notes about formal and informal activities, planned and unplanned activities, symbolic meanings, nonverbal communication, social cues like who talks with whom and who is excluded from conversations, or the identification of power shifts and social exclusion/inclusion; and (5) notes the researcher adds about what should have happened in a setting or interaction that did not happen. There are even researchers who advocate for quantification of interactions, use of language, and other occurrences to look for patterns and implications (Bailey, 2018).

Therefore, these experts support the following list of items to record for observation practices. By incorporating the different levels of verbal and nonverbal interactions at the research site, an observation form must accommodate the following elements:

- Descriptive notes of what you observe
- Notes about what you hear in dialogue
- Reflective notes as a researcher (journal, field notes, recorded notes of nonverbals)
- Demographic details—place, time, setting, locale, other info (weather, lighting, mood)

Other important design considerations include creating a space on the rubric for exact quotes and significant statements; for describing activities chronologically; for providing detailed descriptions with relevant background information to situate the actions or discussions; and to record the date, time, place, and researcher's name for each observation. Also, as noted earlier, observation rubrics can be modified to accommodate the documenting of field notes for targeted ethnographic or cultural studies (see Template 7.2).

Basic Observation Templates

Template 7.1
Observation Rubric for Formal or Informal Settings

Title of Project

TITLE OF STUDY: _____

DATE/TIME/DAY OF THE WEEK: _____

NUMBER OF PARTICIPANTS: _____ SETTING: _____

While observing the setting, the researcher will describe activity related to the following categories:

Relates to:	Individual behaviors	Group behaviors	Nonverbal cues	Conversation topics and threads
Participants				
Setting and use of space/objects				
Types of ongoing activities				
Demographic details				
Researcher reflections				

Note: An exemplar of the observation rubric for formal or informal settings is available in Appendix D.

Template Variations and Challenges

There are several variations on the observation template that may serve different purposes in a qualitative study. Field notes, a specific form of a researcher journal often used in ethnographic studies, include records of observation (verbal, nonverbal), reflections, and cultural symbolism in a blank journal or notebook/log that includes prompts and headings (Template 7.2). A variation for recording field notes in one of these qualitative studies is easily achieved by allowing additional space for descriptive details of the setting (field) and the details of interactions, conversations, demographic details (place, time, locale, weather, mood), and notes about the way people relate to one another and how their conversations evolve or resolve (Angrosino & de Perez, 2000). Regardless of whether your observation activity is formalized or considered a field note exercise for an ethnographic study, the basic template remains the same and is modified by the researcher according to the project's parameters and purpose.

Another variation consists of conducting an observation in order to correlate behaviors, interactions, and nonverbal cues with a theoretical framework that is used for a study. For instance, if you are applying a theory or a group of theories (operationalized in a conceptual framework), you can outline those theoretical components on the left axis of your rubric and observe your population to record the ways in which they reflect those components. Template 7.3 suggests one model for this type of observation activity.

Template 7.2
Observation Log for Ethnographic Field Notes

Title of Project

TITLE OF STUDY: _____

DATE/TIME/DAY OF THE WEEK: _____

NUMBER OF PARTICIPANTS: _____

SETTING: _____

(Continued)

(Continued)

While observing the setting, the researcher will describe activity related to the following categories:

Relates to:	Participants (1)	Topics/ content (2)	Attributes/ characteristics of behavior	Nonverbal cues
Setting, field location				
Demographics (place, time, locale, weather)				
Member characteristics (dress, language/ jargon, rituals, ceremonies, symbolism)				
Group interactions and behaviors, tone/mood				

Template 7.3
Observation Rubric for Conceptual Frameworks

Title of Project

Conceptual Framework	Description of Concept	Observation of interactions, behaviors to theory/conceptual frame elements
Concept	Description	

Concept	Description	
Concept	Description	
Concept	Description	
Connections with other theory(ies) or links with the conceptual framework	Description	

Piloting Observation Rubrics

The key to pretesting an observation rubric is to replicate your study population with a population that will share many of the same attributes, behaviors, and environmental characteristics as your final sample. Use your rubric for a period of time that allows you to sufficiently record, note, and discern a range of interactions that adequately test its functionality. Your pilot test should inform you about the thoroughness of your observation rubric categories, the need to eliminate or add categories, and any notes you might add regarding the tool's connection with your research questions and study's purpose. One thing that is unique to pretesting the observation rubric is that the researcher must allow sufficient time with the pilot population to test the tool(s); much like the prolonged engagement that is required for careful observation, the piloting of the observation rubric requires the same attention to time and engagement.

Transforming Observation Data for Analysis

Kutsche (1998) suggests that preparing observation data for analysis requires two approaches: (1) The researcher should outline the recorded information and create initial typologies to create an overview of the observed realities, followed by (2) organizing the collected data into a narrative, perhaps even a chronological story of events about what occurred. As Miles, Huberman, and Saldana (2014) emphasize, researcher-generated typologies play an important role in preparing voluminous observation data for analysis.

Although not an analysis strategy, per se, member-checking plays an important role in clarifying the observation data before final analysis to allow for those observed to structure the way the scene, cultural context, actions, and conversations are viewed. This collaboration stance, noted as important by Angrosino (2014), reinforces the researcher's dependence on those observed to qualify and codify the data for analysis. Transcriptions derived from recordings, typologies, and observation notes form the basis for the final analysis work. Member-checking further verifies and clarifies the final data set for the application of an analytical approach; however, it is important to view member-checking as a data verification strategy rather than a data collection strategy. Once the raw data have been reduced, the researcher must select the appropriate analytic strategy that best represents the research design connected to the study.

HIGHLIGHTS

Tools for Observation	Observation is the systematic, intensive data collection process for capturing the behaviors, interactions, and social/cultural context of individuals or groups in a setting.
	Observers may be participant-observers or nonparticipant observers, depending on the study; the same rubric may be modified for either approach.
	Observers must possess skills for recording and noting extensive details in a unobtrusive, nonjudgmental manner and must be good at dissecting patterns, meanings, and interactions.
	Observation rubrics are designed to capture the comprehensive interactions of individuals; rubrics are also used in ethnographic field note observations and for conceptual framework analysis.
	Piloting observation rubrics must be conducted on individuals or groups that are similar to but not consisting of the target population, with sufficient time in the field.
	Preparing observation data for analysis requires that data are organized into researcher-created typologies to prepare for analysis using different strategies employed to interpret words, behaviors, and physical context.

CHAPTER 8

Document and Artifact Analysis Tools

> *Document analysis is a systematic procedure for reviewing or evaluating documents to . . . elicit meaning, gain understanding . . . in qualitative research.*
>
> (Bowen, 2009, p. 27)

> *The study of material culture is thus of importance for qualitative researchers who wish to explore multiple and conflicting voices, differing and interacting interpretations . . . and are necessary for most social constructs.*
>
> (Hodder, 2003, p. 159)

Documents and Artifacts Defined

As humans, we create a trail of evidence for others to see, touch, read, and interpret. The very nature of society is to create and generate these points of evidence, which can be captured in numerous formats. There is a core of commonly accepted types of documents and artifacts (sometimes known as material culture) that qualitative researchers can use to ground, supplement, or enhance a study. To use these data sources appropriately, tools must be designed to help the researchers inventory, describe, and assess their value in the context of their study.

Documents and artifacts are, therefore, the readable, tactile, observable, and tangible evidence at a research site. They may exist already (extant), or they may be created by the participants at the request of the researcher (generated); they may even be created by the researcher in tandem with participants (photo voice, windshield surveys). While documents are typically in written or virtual formats, artifacts are typically tactile in the form of artwork, furniture, buildings, photographs, video or audio recordings, material culture objects, signs and symbols, and other tangible objects that represent a research site or its people.

Table 8.1 Document and Artifact Types

Public/Institutional Records Policy manuals, handbooks, official transcripts and course records, institutional self-study reports, program review reports, meeting minutes for standing committees, institutional plans, public records, annual reports, advertisements, radio and television scripts, programs from events or services, maps, charts, press releases, brochures, grant applications, funding proposals, attendance registers, books, publications, survey data, research reports

Personal Records Personal letters, diaries, journals, blogs and online posts, e-mails, calendar notations, personal memos, scrapbooks, photo albums, photos generated by photovoice or windshield documentation, family Bibles

Cultural/Physical Objects Public and private art, signage, posters, sculpture, objects used for ceremony or display, and other forms of cultural symbolism, architecture, and public sculpture

(Bowen, 2009; O'Leary, 2014; Patton, 2015)

Documents comprise a wide range of items, including archival materials, meeting minutes, meeting agendas, meeting presentations, reports for internal and external audiences, institutional plans, institutional self-study reports, and executive summaries about a project or process (Bowen, 2009; Hughes & Goodwin, 2014; Prior, 2014). Artifacts comprise a wide range of items as well, such as physical items from a research site, video and audiovisual recordings, art and material culture objects that are symbols, representations of culture, and other cultural artifacts that stand for something related to the culture under study (Alvesson, 2002; Geertz, 1973; Tinkler, 2013). A more complete list of the range of document and artifact types can be found in Table 8.1.

Document and Artifact Applications

In many studies, documents and artifacts comprise a supplementary data source, and they are frequently used in qualitative designs. Denzin (1970) stresses that document analysis is often used to triangulate with other forms of qualitative data in a single study. In other instances, however, documents and artifacts play a central role, such as in ethnographic, case study, historical, and phenomenological projects (Angers & Machtmes, 2005; Bowen, 2009; Merriam, 2002; Stake, 1995).

The corroboration of stories, narratives, conversations, and shared experiences with tactile, tangible evidence in the form of documents, writings, publications, notes, and material culture objects form the central focus of qualitative explorations.

These distinctions lead to the specific purposes and applications of documents and artifacts in a qualitative project. Bowen (2009) provides an excellent justification of five distinct applications for documentary material; by extension, we can also use these reasons to support the inclusion of artifacts. These justifications consist of using these data to provide context for a qualitative study; for prompting additional questions and observations that need to be probed in a study; for providing supplementary data; for tracking change, development, and evolution of a site or program or phenomenon over time; and, finally, to corroborate evidence with other data in the same study (pp. 29–30). As Angrosino and de Perez (2000) note, evidence that diverges from the primary findings warrants further investigation; the researcher uses these comparisons and reinforcements to strengthen the study and create greater confidence in the interpretations.

Finally, there remains the question of how many pieces of evidence are sufficient within a single study. This question is important to address on several levels. First, the quality of the data always overrides the quantity, and this is just as true when collecting documentation and material objects. Researchers must align their evidence with their research purpose and ensure that what they gather, obtain, and scrutinize fits within the scope of their study. Second, finding evidence that is sufficiently complete, detailed, relevant, and accessible is essential. Locating numerous documents and objects may seem satisfying until you realize they are incomplete, lack substance, or do not answer your research questions, corroborate with your data, or provide new insights in any meaningful way. Thus, the answer to how many documents and artifacts are required for a study can only be answered within the parameters of the study's requirements and the application of data saturation and verity.

Requisite Skills of Document and Artifact Recorders

Researchers require specific skills to identify and analyze documents and artifacts. Careful preparation and organization prior to collecting evidence is a skill that a researcher cannot overlook; making a comprehensive list of all the documents and objects that will support your study requires thoroughness, attention to detail, and a keen sense of where to locate and access your evidence. O'Leary (2014) proposes a planning and discernment process to prepare for collecting documents and artifacts for a study. The following is a modified list of sequenced steps, based on her proposal and combined with Bowen's (2009) and Patton's (2015) recommendations:

1. Create a preliminary list of all relevant documents, artifacts
2. Make a plan for how you will access the documents, artifacts
3. Develop necessary research skills for gathering, interpreting evidence
4. Acknowledge access and bias issues, and make plans accordingly
5. Consider and address ethical issues, protection of institutional/ individual identity
6. Outline alternate options for documents and artifacts if initial plans meet with obstacles; confirm those plans if your original plans for evidence fail

Patience and endurance comprise two other important qualities, since reading and interpreting large sets of data requires time, concentration, and an ability to synthesize material for patterns, themes, and connections with other data in a study. Lastly, a researcher using documents and artifacts to support a project needs to monitor his or her own bias or unconscious selectivity when choosing or reacting to these data. While reflexivity may be applied differently when analyzing documents and artifacts, there is still a concern for how the researcher interprets, understands, applies, and even judges the data under examination (Alvesson, 2002; Bailey, 2018).

Document and Artifact Types

Bowen (2009) categorizes documents into three main groupings: (1) public records, which constitute official, formalized records of an organization; (2) personal documents, consisting of any type of note or correspondence originating from an individual; and (3) physical evidence and objects. The repositories for these items exist in numerous locations, to include written documents in offices, archives, libraries, museums, and other archival repositories. Often, the items can only be obtained through permission of the gatekeepers or similar individuals in authority. Other times, the researcher can access any or all of the items through public sources, including the Internet.

Denzin and Lincoln (2003) refer to documents and artifacts as mute evidence (p. 50) and stress the nature of the researcher's interaction with these resources. They comprise data that must be interpreted by the researcher, and this must happen within the context of the study's focus. Patton (2015) offers a comprehensive inventory of the types of documents and artifacts that researchers might consider for their studies, breaking these types into six major categories: (1) individuals/families, (2) communities, (3) nonprofit groups, (4) programs, (5) Internet groups, and (6) government units (p. 378). Other scholars use different typologies to organize the types of documents and artifacts applicable to a study. Bowen (2009) provides a longer list, with greater specificity to include, but it overlaps with Patton (2015) in many ways (pp. 27–28).

Getting Started With a Basic Template

General Design Considerations

Researchers use rubrics with self-created categories to record evidence generated from compiled documents and artifacts (researcher-generated typologies). These typologies originate from the research literature (inductive) and from the participant's perspectives (in vivo). Additionally, as noted earlier, there is also the distinction between extant documents or artifacts, and those documents or artifacts created at the request of the researcher. A category should exist on a rubric if there is such a distinction in the boundaries of the study.

Regardless, there are several main categories that should be included in every rubric to collect and analyze documents and artifacts. Bowen (2009) and Patton (2015) discuss the categories that a researcher should include, such as where the item is stored or located; the author or creator of the item; the original purpose of the document; and when the item was created. These are the essential properties that allow the researcher to corroborate the evidence with other data and with the study's objectives. Finally, the goal of interpreting and analyzing these data is to identify convergence and divergence from the other data in order to understand where gaps in the narratives, stories, impressions, and perceptions of a phenomenon occur.

Basic Template

Template 8.1
Combined Document/Artifact Rubric

Title of Project

Document/artifact	Location/source and author/creator	Original purpose of item	Date created	Consistent with findings	Divergence from findings
Doc-art 1					
Doc-art 2					
Doc-art 3					
Doc-art 4					
Doc-art 5					

Note: An exemplar of the combined document/artifact rubric is available in Appendix E.

Template Variations and Challenges

One variation on the basic rubric is found in an artifact rubric designed for the sole purpose of collecting and reviewing material culture, objects, and/or artifacts from a research site. This type of rubric may make the analysis of large volumes of artifact data easier to interpret if separated from documentary evidence. Another rubric variation is a worksheet that assists with the researcher's reflexivity, enabling his or her interpretation of artifacts as a separate exercise (Template 8.3). Alvesson (2002) suggests that the study of material culture often leads researchers to confuse the original meaning of objects with their implied meaning within a specific study. By creating a supplementary rubric that allows solely for the researcher's reflection on how the artifact, object, or other form of material culture is contextualized and interpreted may offset possible researcher-based bias or misinterpretation of the data. As is the case in many studies, several rubrics may be combined to assist the researcher in the data collection process; there is rarely a single data source in a qualitative study, and many tools are required to support the process.

A third variation, especially useful with documentary evidence, is the program evaluation rubric (Template 8.4). In this variation, researchers connect program documentation and its corroboration with primary data sources; if documentary evidence supports primary findings, it strengthens those findings. If documentary evidence contradicts primary findings, it warrants further exploration and additional sources of evidence to explain the gaps in the findings.

While variations in design may be limited for document and artifact rubrics, there are additional categories or questions that a researcher may incorporate into a rubric. For instance, a researcher may find particular value in noting the audience for whom a document or artifact was created; questions that arise from reviewing the document or artifact that may lead to other types of supporting evidence; ideas for further investigation based on the review of the documents or artifacts; additional evidence that a researcher may ask study participants to produce or create for review; and, finally, researcher reflections on possible biases inherent in the review of the evidence. The basic rubric can easily be modified to include or substitute these categories with the categories in the basic design.

Template 8.2
Artifact Rubric for Objects/Tangible Evidence

Title of Project

Artifact	Artwork	Signage	Architecture/ use of physical space	Tactile symbols of culture	Videos, recordings
Artifact 1					
Artifact 2					
Artifact 3					
Artifact 4					
Artifact 5					

Template 8.3
Artifact Rubric for Researcher Interpretations

Title of Project

Artifact	Connection to primary data	Researcher's interpretation
Artwork/public art		
Signage		
Architecture		
Use of physical space		
Tactile symbols of culture		

Template 8.4
Document/Artifact Rubric for Program Assessment/Evaluation

Title of Project

Program Components	Convergence with findings	Divergence from findings

Piloting Document and Artifact Rubrics

Piloting document and artifact rubrics follows the same process as with other qualitative tools, except for one difference: You can actually test your rubrics on the same data sources you plan to include in your study without concern for biasing or compromising your final sample of evidence. Using your preliminary rubrics with a review of a few documents (select a variety and ones that are sufficiently detailed), as well as a few artifacts or objects (select a variety in this case, as well), will quickly reveal whether your rubric(s) include categories and typologies that meet your needs.

Transforming Document and Artifact Data for Analysis

Document analysis and the analysis of artifacts (physical or tactile, tangible objects) is a form of qualitative research in which documents are interpreted by the researcher to give voice and meaning to a topic. To this end, analyzing document and artifact data requires several steps in order to merge the content into broader interpretive categories of meaning to support qualitative findings. Bowen (2009) recommends the combination of content analysis and thematic analysis to generate the most useful results. Organizing data into major categories through content analysis, and then making sense of the content thematically, helps the researcher dissect the numerous words, excerpts, quotations, passages, or narratives into a typology and then into clusters of meaning. Those meaning-segments can then be applied to the other qualitative findings to compare for similarities or differences within the same phenomenon (Denzin, 1970). For content analysis, Krippendorff and Bock's work (2009) are excellent resources; for thematic analysis, Boyatzis's work (1998) is easy to apply and understand. Both approaches require considerable sorting and organizing of data that has been collected across many dimensions, requiring careful attention and sorting. As mentioned earlier, the process of managing and reducing large quantities of open-ended data requires efficient organization. Researcher-typologies are an excellent way to begin the data reduction process for document and artifact data, creating lists and formats that enable the researcher to see the data holistically and to begin the process of interpretation.

HIGHLIGHTS

Document and Artifact Analysis Tools

Documents and artifacts are evidence consisting of archival, extant, and generated items that provide a context for understanding the phenomenon under study.

Documents and artifacts often comprise a secondary or supplementary data source.

Careful preparation, endurance, and scrutiny, with an eye for discerning patterns and connections across data, are important skills for researchers working with these data.

Comprehensive lists of documents and artifacts have been compiled by experts and provide researchers with numerous options to support a study.

Rubrics are used to organize these data into categories, enabling interpretation and corroboration with other data.

The basic rubric design can be modified with variations that are specific to artifact analysis, researcher interpretations and reflexivity, and program evaluation.

Piloting rubrics for these data are easily accomplished with data that will ultimately be used in the study.

Data analysis combines content analysis with thematic analysis to achieve interpretation and corroboration.

CHAPTER 9

Reflective Practice Tools

> *Reflective practice encompasses the process of learning through and from experience towards gaining new insights of self- and/or practice; this involves examining assumptions of everyday practice . . . and being self-aware and critically evaluative.*
>
> (Finlay, 2008, p. 1).

Reflective Practices Defined

Reflection is an important element in the practice of qualitative research, and reflective practices provide researchers with a meaningful way to enhance a qualitative study. Reflective practices also enrich a study by broadening the scope of self-awareness, self-consciousness, and cognitive processing that occurs when any topic is probed from a person's unique experience (Alvesson & Skoldberg, 2010). These practices include the participant's reflections as well as the researcher's reflections; the former is considered a participant's reflective stance, while the latter is considered a researcher's reflexive stance. Essentially, in the domain of qualitative research, reflection consists of stepping back and pushing into the direction of the inquiry so that the researcher can frame his or her own understanding and assumptions and so the participants can do the same.

When soliciting a participant's reflection of any type, the researcher is asking that individual to review, reconsider, process, clarify, or debrief an earlier conversation, interaction, or even a written documentation of an experience. This process requires the researcher to offer an intervention of some kind to help the participant structure that reflection. Conversely, a researcher's reflexivity reflects his or her background, knowledge, bias, methodology, and perspective juxtaposed against a study. Another way to express reflexivity is to say that it represents what the researcher knows about himself or herself and the participants, continuously recorded as a way to offset preconceived notions about the research that might interfere with data analysis and interpretation (Malterud, 2001; Patton, 2015).

Creswell and Poth (2018) add to this explanation by suggesting that researchers should practice reflexivity in order to continuously inform their interpretations about what they are learning from their inquiry (p. 47).

Therefore, to what extent has the researcher worked to neutralize his or her own bias, motivation, or interest as findings are reported? While it is important to embrace bias as an essential component of qualitative research, this role of "researcher as data collection instrument" must be balanced by allowing each participant's voice to override the researcher's assumptions. As noted previously, the goal of qualitative research is not to aim for generalizability but to seek depth, detail, and the multilayered perspective of each participant's experience with the phenomena in question. Therefore, reflective tools must provide a safe space for either the participants or the researchers to consider their feelings, perspectives, and biases regarding experience with a phenomenon.

Reflective Practice Applications

Whether a participant's reflection or a researcher's reflexivity is involved, reflection enhances a study by providing another perspective on the phenomenon in question. These stances of participant or researcher are important ones to consider when integrating reflection into a qualitative project.

While numerous experts highlight the role of the researcher's reflection (reflexivity; Alvesson & Skoldberg, 2010; Patton, 2015), far less information exists regarding the role of participant reflection in qualitative studies. Further, there are very few examples to guide researchers in the design or inclusion of reflective tools in their projects; however, the value of following up with or tracking participants' experiences, journeys, or reflections cannot be understated. Individuals need to process their life experiences. As a researcher probes a participant's perspective, their discourse disrupts or dislodges an experience from the participant's subconscious to their conscious mind (O'Cathain & Thomas, 2004). This shift from an embedded to an acknowledged perception allows an individual to clarify and understand what those experiences mean to him or her. This process takes time, and that timing is unique to each individual (Pennebaker & Seagal, 1999).

The reflective process is one way to capture the emergence of this understanding and interpretation on the part of the individual. As any qualitative researcher knows, uncovering an experience toward a better understanding of the meaning of that experience can be a long journey. As Pennebaker and Seagal (1999) note, an individual's reflections, which result from revealing personal experiences, help that individual clarify meaning for himself or herself and for others. As a data collection strategy in their own right, reflective tools create an opportunity for participants to extend their self-awareness.

While the reflexive practice of the researcher is not the main focus of this discussion, it is still important to understand how researchers may record and process their own reflections during the course of a qualitative study. As Alvesson suggests (2002), and as Patton reinforces, (2015), reflexivity constitutes the practice of a researcher positioning oneself in order to offset inherent biases and assumptions that might otherwise compromise the participants' meaning regarding their experiences or the essence of their lived experiences. This ethical stance is an important element of reflection, one where many researchers find the value of journaling as a critical element in conducting a qualitative project. As Berger (2015) suggests, reflexivity is an important way to ensure quality control in qualitative research by clarifying the researcher's stance in relation to the participants and the focus of the study. Reflexivity is also known as researcher-position, insider-outsider stance, and a self-knowledge strategy; the goal of reflexivity, regardless of the label, is for researchers to understand their role and impact in the creation of knowledge and interpretation for a study (Alvesson & Skoldberg, 2010; Eppley, 2006). For this reason, the topic of researcher-generated reflection is included in this chapter discussion to provide a deeper understanding of the types of tools that both participants and researchers might use in reflective practices. Templates provided for reflection practices are also intended for a researcher to use and modify for reflexive practices.

Requisite Skills of the Reflective Researcher and Participant Reflections

The practice of reflection requires introspection. Creswell and Poth (2018) stress that participant meanings must maintain and uphold the integrity of a participant's perspective on the phenomenon in question. In order to avoid obscuring the meaning that a researcher may ascribe to participant stories, it is critical to solicit the participants directly regarding their experiences. In this way, reflection is best defined as an attempt to promote transparency in participant perceptions (Ortlipp, 2008). Multiple perspectives from the same participant enrich and substantiate the ways in which a phenomenon is grounded. Therefore, the reflective researcher must employ skills that elicit these reflections from participants without diffusing or diluting their meaning.

These skills require the same sensitivity that other forms of qualitative data collection require, but the reflective researcher is often at a distance from their participants; in this way, a researcher must demonstrate the necessary skills of being trustworthy, empathic, and nonjudgmental in order to obtain transparent reflective data from participants. Even after an initial interaction with participants, if a researcher is deemed to be untrustworthy or insincere, participants will either refuse to complete follow-up reflective questionnaires or will only provide superficial information (Agee, 2009; Wang, 2013).

Reflective Format Types

A number of formats are designated to capture reflective data in a qualitative study. The most common formats include reflective questionnaires, journals, and diaries. Each of these formats is used for different reasons and in different types of qualitative studies.

Reflective questionnaires. The reflective questionnaire is typically a participant-generated response device. This type of questionnaire consists of an open-ended device that captures participant thoughts and perspectives, often at the end of a research process, and as a form of debriefing or follow-up (Brody, Gluck, & Aragon, 2000). These questionnaires are used in a wide variety of qualitative studies. Reflective questionnaires are far more than open-ended questions positioned at the end of a structured survey instrument; similarly, they are not a substitute for member-checking or other trustworthiness strategies.

Reflective questionnaires are an effective method for closing a conversation with participants, thereby securing any remaining insights that can enhance a study. As humans, we are constantly searching for meaning, and we synthesize our life experiences in unique ways (Brinkmann & Kvale, 2014). As people consider and reflect on their personal stories, they gradually change how they perceive those experiences; once an experience assumes structure and meaning, the exercise allows for closure and resolution. Researchers who are able to capture these residual comments, thoughts, observations, and ideas can be assured that they are receiving substantive reflections resulting from the initial discourse (Reid, Flowers, & Larkin, 2005). In these ways, reflective questionnaires generate new understandings and new perspectives that enrich a qualitative study.

As Gill, Stewart, Treasure, and Chadwick (2008) stress, "this can often lead to the discovery of new, unanticipated information . . . respondents should be . . . debriefed about the study after the interview has finished" (p. 293).

Journals and diaries. Both types of repositories capture the participants' or the researchers' ongoing recording of their perspectives, experiences, observations, and interactions with the research site and research participants. Although similar in format and reflective in scope, journals tend to focus more on the researchers' observations of the research setting while diaries may focus more on the personal journey and reflections of the participants. This is not universally true, but the distinction is important when considering which tool to use and who is generating the data. Either way, journal and diary entries allow individuals to freely structure and document their journey or experience as they choose. These documentations may be as frequent or as episodic as the individual determines, but the entries are open-ended, unstructured, and conversational in nature. Individuals are likely to include words, pictures, drawings, notations, symbols, or any other form of language that expresses their feelings and perspectives. These devices also serve as first-hand accounts of a specific experience, and the resulting data generates a story that evolves and solidifies over time (Bernard, 1988; de Laine, 2000; Ortlipp, 2008).

Table 9.1 outlines the various research designs that use reflective tools and their purposes (Alaszewski, 2006; Bailey, 2018).

Table 9.1 Reflective Practice Tools, Purposes, & Qualitative Designs

Reflective Tools	Purpose	Qualitative Design
Reflective questionnaires	Enhancement & follow-up strategies to capture participant perspectives as a culminating phase of a study	Descriptive/interpretive Phenomenological
Journals/Diaries	Self-generated records of an individual's journey/experience that is written, drawn, or diagrammed on a daily, regular, or episodic basis to reflect personal perspectives, views, and biases	Descriptive/interpretive Phenomenological Ethnographic Narrative studies Grounded Theory

Getting Started With the Basic Templates

General Design Considerations

Participant perspectives and stories are at the heart of any qualitative design, frequently augmenting previously collected data. These succinct questionnaires are best described as follow-up devices, comprised of one or two open-ended questions that encourage participant reflection, represented in several common formats. The most common application of this tool consists of e-mailing, texting, or mailing an open-ended questionnaire to participants and asking for a prompt response (the initial response may be diluted if too much time is allowed to lapse). While simple in design, reflective questionnaires contribute to a study in important ways and confirm findings generated from the same study.

Researchers have numerous options to consider when designing reflective questionnaires. The basic reflective questionnaire includes one of a few simple approaches:

Single topic questionnaires. These formats extend the essential question(s) from interviews or focus groups by asking if there is anything else the participant wishes to share regarding their experience.

Free association questionnaires. These formats solicit feedback linked with specific concepts that emerged during the initial interviews, focus groups, conversations, or interactions.

Scenario questionnaires. These formats ask participants to indicate their response to a specific situation, thereby highlighting their beliefs and reinforcing their responses resulting from interviews or focus groups.

The most common method for administering a reflective questionnaire is to send participants a link to an online survey platform (e.g., SurveyMonkey, Zonka, SurveyGizmo) to facilitate the ease of completion. Creating a brief questionnaire in a survey program makes the access to and completion of the questionnaire a seamless exercise for participants; conversely, data collected through one of these platforms ensures efficient data retrieval, reduction, and analysis for the researcher. Typically, the reflective questionnaire link is sent to participants within 24 to 36 hours after the initial interview or interaction. This allows the participant sufficient time to process the original conversation and process their recollections; the delay and the distance allows the participant to clarify, extend, modify, or add to his or her original comments and insights.

When using a survey platform to create a reflective questionnaire, there are two ways to approach the design process. For instruments designed to collect narrative or scenario-based responses, developing a sequence of open-ended text/comment boxes is the best option. As you create these questions, it is important to allow for the maximum number of words/characters for each text/comment box; you can indicate that participants have a specific number of characters or even an unlimited number of characters within which to respond. Conversely, when creating the free association questionnaire, survey platforms allow for the creation of questions that allow participants to enter finite responses, limited to short answers in boxes. These question options can be found in the dropdown boxes that appear in these programs under "question construction." In all three types of reflective questionnaire designs, you want to leave room for your participants to enter any final or concluding thoughts by providing an open-ended comment box at the conclusion of your questionnaire or prompt.

Researchers may also create a Word document and attach that document to an e-mail, sent to all participants with a request that they complete and return the questionnaire. This option is less than ideal since it involves extra steps to download, complete, re-save, and attach these documents in order to ensure a prompt reply to the sender. Since the goal is to make the response exchange as efficient and as relatable as possible for the participant, the online survey platform remains one of the best options for administering reflective questionnaires.

Other options for creating these tools include interactive applications such as QuickTag, BeSocrative, or even Google.docs, which enable participants to respond quickly and effectively. The challenges with these online or interactive formats is manifested in the data retrieval, management, and analysis process. Reflective data must be whole and readable in order to be adequately analyzed. Many of these ease-of-use platforms compromise the essential analysis process. These challenges likewise make the use of text messages or other social media platforms less than ideal (e.g., PollEverywhere, Kahoot!, PigeonHole); the data cannot be accessed in a workable format for analysis. Therefore, when designing and identifying the appropriate application for delivering a reflective questionnaire, the researcher must consider all subsequent phases of data management.

Basic Reflective Questionnaire Templates

Template 9.1
Single Topic Reflective Questionnaire

This template provides the participant with another chance to reflect on, process, and disclose final thoughts regarding the topic of inquiry probed in the prior phase. Beginning with a brief introduction, with simple guidelines and a response deadline, this form is intended to elicit direct, candid, and detailed feedback. This is the simplest format for reflective questions and generally reflects a topic covered previously.

Title of Project

Thank you, again, for your willingness to meet with me regarding (insert title of the study). I would like to ask you one final question to follow up on our discussion.

QUESTION: Insert a single question that directly relates to the topic of your prior meeting.

Example: Now that you have had a chance to reflect further on our discussion regarding (insert topic), is there anything else you wish to share regarding your experience/perceptions?

Please feel free to share any thoughts, details, examples, or experiences that you believe will add to our discussion on this topic. Your insights and your perceptions are important and will add richness to this study's findings.

I would appreciate your response within the next week, by (insert date). Please feel free to contact me if you have any questions or wish to talk further.

Thank you,

Signature/researcher's contact information

_____

_____

Note: An exemplar of the single topic reflective questionnaire is available in Appendix F.

Template 9.2
Free Association Reflective Questionnaire

This template asks participants to respond to a cluster of words/phrases related to the research topic. Beginning with a brief introduction, simple instructions, and a response deadline, this format is designed to produce generative and creative reflection. The exercise helps participants communicate deeper emotional responses within the context of their prior comments. This strategy is particularly effective when used in phenomenological designs, as it probes the individuals to tap the underlying sentiments, perceptions, and triggers that come with phenomenological inquiry.

Title of Project

Thank you, again, for your willingness to meet with me regarding (insert title of the study). I would like to ask you respond to the following words/phrases as a way to follow up on our earlier discussion.

FREE ASSOCIATION QUESTION:

Example: What comes to mind when you read the following words or phrases? Provide one or two responses to each item, as they immediately come to mind. Your first intuitive impulse is typically the most representative of your experience.

Word/Phrase _____
Word/Phrase _____
Word/Phrase _____
Word/Phrase _____

Your insights are important and will add richness to this study's findings. I would appreciate your response within the next week, by (insert date). Please feel free to contact me if you have any questions or wish to talk further.

Thank you,

Signature/researcher's contact information

Template 9.3
Scenario Reflective Questionnaire

This template prompts participants to indicate how they would react or behave in the designated circumstances, extending prior discussions with the researcher. Beginning with a brief introduction, with simple instructions and a response deadline, this format challenges the participants to synthesize what they have said previously by revealing their assumptions and beliefs. The questions should be carefully constructed and succinct in order to ease the participants into the response process.

Title of Project

Thank you, again, for your willingness to meet with me regarding (insert title of the study). I would like to ask you respond to the following scenario as a way to follow up on our earlier discussion.

SCENARIO QUESTION

Example: Please consider the following scenario, and describe how you would respond. Please feel free to share as much detail and explanation as possible, with examples, experiences, or insights that support your description.

(Insert scenario)

Your insights are important and will add richness to this study's findings.

I would appreciate your response within the next week, by (insert date). Please feel free to contact me if you have any questions or wish to talk further.

Thank you,

Signature/researcher's contact information

Template 9.4
Reflective Journal and Diary Logs/Notebooks

Many journals and diaries are nothing more than blank notebooks in which individuals may freely record their impressions, observations, thoughts, and aspirations. Some researchers find that providing a single prompt for individuals serves as a useful catalyst to focus the individuals' thought process when making an entry; in some cases, entries are submitted to a researcher for periodic reading and commenting, while in other cases, the entries are submitted as completed entities, ready for analysis as a final data source. Finally, researchers may use the journals for their own reflections, following the same process that they would ask of a participant; in this way, journals and diaries can be distinguished from researcher-generated or participant-generated documents.

Title of Project

NAME OF PARTICIPANT: _____

NAME OF LOG: _____ DATE: _____

(INSERT PROMPT, IF DESIRED) _____

_____
_____
_____
_____

Template Variations for Reflective Tools

While the formats displayed above are the ones most commonly used in reflective canvassing, each template can be customized for your study's specific focus. The design of all reflective tools should include words or phrases that serve as catalysts that encourage participants to share more of their stories, experiences, or insights. This exercise in transparent disclosure is an important way to close the loop on the participants' perspectives on a phenomenon, experience, or event; the more opportunities you provide for a participants to extend their disclosure, the richer your information. Regardless, just providing a reflective medium for participants, whether it is a blank page or a page with a prompt, signals that their feedback is important to the study's findings.

Piloting Reflective Tools

As with all tools, a pilot test is essential in order to ensure the collection of viable data. To start the pilot process for a follow-up tool, identify a few individuals who resemble your target population and ask them to answer some questions from your primary instrument (such as an interview protocol). This process ensures that the testing of the pilot instrument extends the nature of the original inquiry. Next, test the follow-up questionnaire for clarity, and solicit feedback from the pilot participants regarding the ease of responding to the questions.

Transforming Reflective Data for Analysis

Reflective questionnaires generate words and phrases that allow the researcher to organize, manipulate, and categorize them according to the research question(s). Before delving into a particular content or text analysis strategy, the starting point with this type of data is always data management to data reduction (Billups, 2012). Each type of reflective data requires its own approach.

Reflective questionnaire data. For open-ended narrative text analysis, the key is to reduce the volume of words and phrases to a manageable size. Your choice is to transform these words and phrases into matrix formats or create transcripts that collapse all the raw data into a single file, tangible or virtual. In the case that you prefer to work with paper documents, the process is much like dealing with any other data transcript; make readable copies of the questionnaire response sheets and begin the coding-to-theme progression task. Miles, Huberman, and Saldana (2014) suggest the three-step process of coding for basic analysis activities where data is not too extensive; their descriptive to interpretive to pattern coding strategy is very easy to use.

Journal and diary entry data. By its nature, journal and diary entry data is voluminous. Bailey (2018) points out that the extensive amount of data in the form of words, phrases, and jargon resulting from these sources may seem overwhelming at first. Her step-by-step process for preliminary analysis of narrative/text data is an easily understandable guide, comprised of seven steps, moving from raw data cleaning to coding to theme emergence.

Computer-assisted programs are often used with open-ended questionnaire data, and NVivo is a popular software program for this type of analysis. The choice of whether to preserve the data as paper records or to use a computer-assisted program that facilitates the organization of words/text/stories into categories is a personal one; many researchers resort to these computer programs to make the process more workable. NVivo, HyperRESEARCH, and ATALIS.ti are the programs used by many researchers, but new programs are always released into the market, offering many benefits to the first-time researcher/analyst.

For all types of qualitative data analysis, and especially for reflective data that consists of long passages of narrative, using computer software tools may seem tempting and perhaps even enticing as a short-cut to the work of analysis. It is important, however, to note that computer-based analysis programs do not necessarily reduce the amount of time it takes to analyze data, particularly the data resulting from pages and pages of journal or diary entries. In the end, the researchers must have their own vision of how to organize their data to answer their research question.

HIGHLIGHTS

Reflective Practice Tools	
	Reflective practices allow participants and researchers to review, reconsider, process, debrief, or document the practice of self-awareness, positioning, and exploration about a phenomenon.
	Reflective practices are typically used as a follow-up strategy or as a documented strategy to elicit personal journeys, and experiences from participants or researchers.
	Participant reflection is the intentional strategy to encourage an individual to move from embedded perception to acknowledged reality.
	Researcher reflexivity occurs when the researcher positions himself or herself to process and offset biases and assumptions during a study.
	Reflective skills require insight, intuition, self-awareness, and a capacity for expressing inner feelings and emotions.
	Reflective questionnaires, journals, and diaries are commonly used reflective tools; reflexive practices by researchers may modify these templates for use.
	Reflective questionnaires are comprised of open-ended questions, free word associations, or scenario questions; journals and diaries may include prompts but are also blank logs or notebooks for open-ended recording of impressions.
	Piloting reflective tools requires pretesting instruments with participants who resemble but who will not comprise your final population.
	Preparing reflective data for analysis focuses on interpretation of words, phrases, and other linguistic devices that reflect participant's introspective recall.

CHAPTER 10

Synthesis

The Qualitative Story

> We think metaphorically of qualitative research as an intricate fabric comprising minute threads, many colors, different textures, and various blends of material.
>
> (Creswell & Poth, 2018, p. 41)

A Multifaceted Enterprise

Qualitative research requires the integration and coordination of many different components. Lincoln and Guba (1985) describe qualitative research as naturalistic, emergent, and multidimensional; Creswell and Poth (2018) describe the qualitative study as a complex but well-coordinated patchwork of many elements. In short, the qualitative approach is inherently interpretive, reflecting the rich nuances of the human condition. Thus, a qualitative project requires flexible, adaptable tools in order to capture the tangible and the intuitive aspects of the natural world. Moreover, these tools must be compatible and mutually supportive when they are used in a single project. If qualitative studies depend on multiple perspectives, then it follows that multiple data sets are required. Multiple data sets require multiple devices, all of which must work in tandem to collect extensive data.

There are no absolute boundaries in qualitative research practice. The premise that the right tool connects the research design and the participants with verifiable data suggests that the process is seamless—you should hardly see the breaks between the study's design, who is involved, how the data are gathered, and how the final story is conveyed. Trying to keep each part of the qualitative project discrete, separate, and sequential will only confuse the process; qualitative research is, by its nature, overlapping, fluid, and circular. Even so, the resulting story will be more convincing and more authentic if the procedures are rigorous, carefully planned, and systematically applied.

Therefore, qualitative researchers must acknowledge some working principles to guide their work. These principles support a holistic and integrated qualitative design process; the unique role of the practitioner/researcher; and the need to remain creative, adaptive, thorough, and organized. While these principles may seem contradictory when referring to an open, fluid, interpretive process, they work together for a meaningful outcome.

The Practitioner's Perspective

Qualitative researchers represent many disciplines and assume many roles. Your professional discipline or academic background makes little difference in the actual roots or practice of qualitative research. The health sciences field has pioneered much of the work in the qualitative domain, particularly in areas such as validity, rigor, trustworthiness, reflexive practices, case studies, and ethics. Much of the important work in reflective interviewing and participant journaling is evident in the journal articles from that same field (e.g., *American Journal of Occupational Therapy, Family Practice, Health Services Research, Qualitative Health Research*). Similarly, qualitative research finds its roots in the disciplines of anthropology, cultural studies, psychology, sociology, education, linguistics, ethics, law, and marketing. Many of the foundational texts and resources that support qualitative practices originated in another field of practice; those journals and texts populate the qualitative landscape extensively. If you are searching for information, research articles, and resources to ground your study, expand your perception of what constitutes a research resource to include the numerous experts and scholars in these diverse areas.

Second, your background or your role as a researcher plays an equally important part affecting the design of your study and tools. If you are new to the qualitative domain, the concept of collecting narrative, audible, observable, textual, or tangible evidence may seem daunting enough without the added concern of how to design tools to capture each distinct data type. If you are a graduate student, if you are a faculty member teaching qualitative research methods, if you are a veteran researcher, or if you are new to qualitative research after spending years working on quantitative projects, your approach to designing and developing qualitative tools will vary.

As a graduate student who is working on his or her thesis or dissertation, your primary goal may be one of expediency. You want to find the best and most direct path to creating tools to assist you with your research project. You need a hands-on text, practical examples, and general guidance to help you complete a rigorous and well-designed project. As a faculty member teaching others about qualitative research methods, you need resources and exemplars that will guide your students and help them understand the range of options and approaches in the qualitative arena. As a researcher, whether you are a veteran, new to qualitative research, or a team member on a project, you need a resource you can access that will help you determine the best way to collect qualitative data that is neither too simplistic nor too complicated and can be easily merged with other parts of the study. Regardless of your role, a resource that provides templates and examples of qualitative tools, explanations of how these tools can be used in a study, and the rationale for the variations among and across the different types of tools is essential to your work.

The Qualitative Design Process

This text seeks to assist qualitative researchers with the knowledge and the tools to conduct reputable qualitative research projects. Based on the assertion that no study is viable without viable data, it follows that all data collection depends on employing the appropriate data collection tools. However, the concept that qualitative research projects require viable tools has its proponents and detractors; some scholars suggest that instruments, tools, or predesigned devices have no place in a qualitative study. Yet, I contend that no study can succeed without effective data collection tools, and this includes qualitative studies. If a research tool is defined as a device for achieving careful and exact work in order to accomplish something of value, then interview protocols, focus group moderator guides, observation rubrics, or reflective questionnaires may all be classified as research tools designed to facilitate data collection.

Further, each part of a qualitative project is interconnected and interdependent. The research purpose and research question(s) determine the research design; the research design determines the role of the participants; the relationship between the participants and the research design leads to the type of tool(s) needed to collect the data. In most qualitative studies, a combination of tools must be designed for a unified purpose, as well as to ensure triangulation and the verity of the findings. For instance, in an ethnographic study, interview protocols, observation rubrics, journals, diaries, field note logbooks, document and artifact rubrics, and conversational/discourse rubrics may all be employed. The design of one tool within such a study requires careful coordination with all the other tools included in the same study; the flow between and among the other tools is essential to the integrity of the data collection procedures. Therefore, a researcher must assume a holistic design approach when creating a set of qualitative tools. When all these essential pieces converge, the researcher is empowered to fully explore the phenomenon under study and effectively tell the stories from the participants' unique perspectives.

To guide this complex work, it is essential to develop a research blueprint for your qualitative study that connects all the parts from the start,

Table 10.1 The Qualitative Design Process

Research Design	Data Collection Strategies	Data Collection Tools
Descriptive/ Interpretive	Interviews, dyads/triads	Interview protocols
	Focus groups	Moderator guides
	Documents	Document rubrics
	Observation	Observation rubrics
	Reflections	Questionnaires
Phenomenological	Depth interviews	Interview protocols
	Reflections	Journals, questionnaires
	Documents	Document rubrics
Ethnographic	Depth interviews	Interview protocols
	Documents	Document rubrics
	Artifacts	Artifact rubrics
	Observation	Observation rubrics
	Reflections	Discourse/conversational tools
		Journals–informants
		Journals–researcher
		Field notes

(Continued)

Table 10.1 (Continued)

Research Design	Data Collection Strategies	Data Collection Tools
Narrative	Depth interviews Reflections	Interview protocols (life history, bio) Questionnaires Journals–informant Journals–researcher
Case Study	Interviews, dyads Focus groups Documents Artifacts Observations	Interview protocols Moderator guides Document rubrics Artifact rubrics Observation rubrics Discourse/conversational tools
Grounded Theory	Interviews Documents Reflections	Interview protocols Document rubrics Questionnaires
Historical	Interviews Documents Artifacts	Interview protocols Document rubrics Artifact rubrics

rather than as separate pieces that are created as each phase of the study unfolds. Table 10.1 revisits Table 2.3 to emphasize these connections:

As Table 10.1 indicates, the research design informs the data collection strategies, which, in turn, dictate the data collection tools. This design process requires careful planning and coordination.

A Plan for Action

As noted, a qualitative project is a multilayered effort. Wise researchers will create systems to organize their work and the phases of a study. Many researchers create matrices, flow charts, diagrams, or conceptual maps to track their work and illustrate the connections between the data

Table 10.2 The Qualitative Data Collection Plan (SAMPLE)

Data Sources	Date/Time	IRB	Interviews	Focus Groups	Documents/Artifacts	Observations
Participants						
Pilot interview						
Interviewees, Phase I						
Interviewees, Phase II						
Pilot focus group						
Focus Group 1, Phase III						
Focus Group 2, Phase III						
Focus Group 3, Phase III						
Observations						
Pilot rubric						
Participants observed						
Documents, Artifacts						
Pilot rubric						
Archival, extant, Phase IV						
Created evidence, Phase IV						
Material culture, Phase IV						

and the tools. Therefore, the easiest way to begin a qualitative project is to develop a working outline of the data collection phases and coordinate those phases with the participants, sites, or other evidence, and the timeline for accessing those sources. A sample data collection plan is outlined in Table 10.2.

It is very easy to lose your way while conducting a qualitative study; many data collection activities occur concurrently. Some phases of a study are informed by a prior phase, and the timing is critical to developing the next type of tool or device. Make a plan, and organize your project carefully. You can revise your basic matrix or chart in any way that makes sense to you, but developing a system to track your study phases, participants, and tools will contribute to more efficient procedures.

Recommendations

The following recommendations reflect some of the ambiguity, some of the contradictions, and some of the unanswered questions you may face as a qualitative researcher designing your own qualitative tools. My recommendations are meant to be practical and are offered to help you frame your work within the current state of the qualitative research movement.

The templates are just the starting point . . . it is up to you after that! This text provides templates to help researchers begin their design process with basic templates; after that, the researchers must decide how those templates can be customized to reflect the purposes of their particular study. The templates, then, are offered as starting points on the research journey. If you have clearly delineated your research purpose and objective, you can leverage the templates in numerous ways.

To transform templates into tools specific to your study's purpose, you will need to read as much as you can about your selected research design and ground yourself in the work that has already been accomplished in your topic area. This information will help you design the appropriate tools, but you must also keep an open mind and trust your instincts about what you need and how you want to develop those tools. I guarantee there will be multiple viewpoints on any given design and any given tool; keep an open mind and have confidence in your own instincts.

Be creative in your approach. Not only should you trust in your own instincts and abilities when it comes to designing your qualitative instruments, but you should also believe in the power of creativity and experimentation in this process. The templates in this text are basic, and none of them should be used without adaptation or modification. Any change you make as the principal investigator depends on your research goals and your research participants. Experiment with various approaches to instrument design; by piloting every tool you intend to use in a study, you will have the chance to refine and correct any flaws or inconsistencies. The interpretive approach essential to a successful qualitative study begins with your interpretive approach to instrument design. Exemplars for select templates are provided in the Appendix to provide a basic framework, but the final judgment regarding the applicability of any tool rests with the researcher.

Read extensively and entertain the debate among qualitative scholars. The debate in and around the qualitative movement is extensive, contradictory, and unending. The more deeply I become immersed in the world of qualitative research, the more I acknowledge the variations in qualitative definitions, applications, labels, origins, and foundations. If I read 20 texts on a particular research design, I may walk away with a profound sense of the commonalities across those texts, but I will also walk away with 20 different interpretations of how, why, and when to use the qualitative approach. If you are uncomfortable with ambiguity and blurred lines, qualitative research is not for you! If you can absorb the opposing viewpoints and find the common ground in all of these perspectives, you will enrich your understanding of the qualitative approach. As I see it, the differences among the experts and scholars merely reflect the interpretive nature of this research approach; there are multiple viewpoints, multiple realities, and multiple applications. It can enhance your understanding or unnerve you—it is your choice.

Piloting your tools and your processes guarantees a rigorous study. As mentioned earlier in this text, every protocol, guide, rubric, log, and open-ended questionnaire used in a project should be pretested. Beyond the step-by-step procedures for pretests, the practice of piloting should be embedded in all of your studies. Determining whether a qualitative tool is well-designed, and whether it will collect usable data, depends on three fundamental things: (1) your attention to the flow, sequence, clarity, and purpose of each question, probe, and category in each of your tools—what works, what does not work; (2) asking pilot participants (when appropriate) to offer their feedback on the tools you have tested with them and comparing that feedback with your own observations; and (3) conducting preliminary data analysis on your pilot data to determine whether your data are viable for analysis, and later, for interpretation. There is no sense in collecting data you cannot use.

If you routinely apply these overarching strategies to piloting your instruments, in conjunction with the specific piloting procedures outlined in the chapters, you will avoid problems later on when a course correction is problematic or, worse, not possible at all.

Conclusion

We live qualitatively. During the course of our lives, we speak, hear, touch, see, and perceive the world around us, engaging with people and their stories, much like the intricate fabric of threads, textures, and materials noted by Creswell and Poth (2018). The act of listening, conversing, observing, recording, noting, reflecting, and probing requires sensitivity and patience as the phenomenon emerges, in its many layers and from the different viewpoints of the individuals sharing those experiences. The qualitative researcher must acknowledge the duality of letting the stories take shape while guiding the stories into something meaningful that others can read or hear or understand. Thus, being a qualitative researcher does not mean you cannot or should not appreciate order, precision, or control over your procedures; rather, the qualitative approach requires a researcher to allow the inductive process to guide rather than dictate the final outcome. This volume contributes to the exploration, discovery, and unfolding of the qualitative story by offering practical tools for the researcher. I hope you will appreciate its value as you continue your qualitative quest.

Appendices
A Case Study of Department X

The following exemplars are based on a case study project regarding the formation and development of an academic department within a large university. This study was designed as a single case study with the academic department designated as the single case, bounded by the timeframe of its establishment through the five-year mark. The study explored best practices regarding the creation of a new academic department that housed several new programs and the relationship of that department to the larger institution. Issues of community, culture, leadership, and sustainability were included in the exploration of how academic leaders might plan for and implement new programs on their own campuses.

Six qualitative tools were used in this project, designed to capture the perspectives of stakeholders and campus leaders. Each of the tools illustrates how a basic template can be customized to match the research objectives and research question(s) for a particular study; in this instance, a case study was chosen for these examples because of the nature of the approach. Case study research involves numerous data sources requiring extensive data collection; thus, the number and range of tools served the purpose of showcasing these exemplars and illustrating how a set of data collection tools can be coordinated with a single focus and comparable design. The tools and data collection are organized by study phases.

The research questions guiding this study were:

RQ1 (overarching question): How do stakeholders describe the formation and development of a new academic department in the context of its position within the larger university, its cultural formation and grounding, its leadership and advocacy, and its operational effectiveness?

>RQ1a: How do stakeholders perceive the department's relationship to other academic departments and in the larger university context?

>RQ1b: How do stakeholders perceive the emergent department culture as it conforms to and departs from the University's dominant culture?

RQ1c: How do stakeholders describe the role of leadership in the development and growth of the department?

RQ1d: How do stakeholders distinguish the department's strengths, challenges, and opportunities for innovation and growth?

RQ1e: How do stakeholders envision the department's future over the next five years?

Appendix A Exemplar
Unstructured Interview Protocol

Title of Project

DATE: _____ TIME & PLACE: _____

INTERVIEWER: Dr. X   INTERVIEWEE: Program Director & Founder

OTHER: _____

Pre-Interview Information & Procedures

<u>Introductions:</u> Researcher introduces himself or herself, reviews process for session, how long interview will last, and general format for questions

<u>Study purpose and applications:</u> Researcher reviews study's purpose and uses of the findings, including how the findings will be reported and shared

<u>Consent forms, approvals:</u> Informed consent forms distributed to participants, signatures secured, assurance of privacy/confidentiality/anonymity as appropriate, protection of the participant assurances reviewed, questions answered; note that the interview will be recorded and obtain permission for that, as well

<u>Treatment of data:</u> Researcher indicates how data will be managed, secured, and disposed of after a specific time period

<u>Other questions or concerns?</u> Other issues are discussed prior to beginning the interview session

Opening the Interview Session

Opening question: Use the initial question to introduce your topic and to establish a rapport with your participant.

Q1: Opening question

> Tell me about the way the vision for this new department originated and how you made that vision a reality?

Key Interview Questions

The central portion of this interview form consists of one or two questions that set the stage for the conversational mode you are facilitating; add probes that can be used sparingly during this conversation.

(Continued)

(Continued)

Q2. Content Question: *Tell me about the relationships, politics, and institutional dynamics that you navigated to establish this department and launch its programs.*

Probes: Can you give me a specific example?

Q3. Content Question: (alternate or extension question) How would you describe the vision for the next five years for the department? Its challenges? Opportunities? Allies? Foes?

Probes: Is there a story(ies) associated with that incident?

Q4. Content Question: (alternate or extension question) Describe a typical week in the department now. How might a typical week look five years from now?

Probes: Can you explain why you feel things will be so different (or the same) five years from now?

Concluding the Interview

Transition to the end of your interview session with a question that allows the participant a chance to debrief or communicate any final thoughts, clarifications, or comments that still need to be shared. A single open-ended question, posed by the researcher, is the best way to capture these final sentiments or thoughts.

Q6. Concluding Question:

Researcher script: *To obtain your final thoughts, is there anything else you would like to tell me or share with me regarding today's topic?*

_____

_____

Thank You and Follow-Up Reminder

Researcher Script: *Thank you for your time and your insights on (insert topic). I will follow-up with you in a few days to (choose one or more of the following) (1) ask you to complete a reflective questionnaire, (2) complete a member-checking exercise to verify my notes of our session, or (3) ask you a few questions for clarification.*

Appendix B Exemplar
Semistructured Interview Protocol

Title of Project

DATE: _____ TIME & PLACE: _____

INTERVIEWER: Dr. X INTERVIEWEE: Faculty Member #1

OTHER: _____

Pre-Interview Information & Procedures

<u>Introductions:</u> *Researcher introduces himself or herself, reviews process for session, how long interview will last, and general format for questions*

<u>Study purpose and applications:</u> *Researcher reviews study's purpose and uses of the findings, including how the findings will be reported and shared*

<u>Consent forms, approvals:</u> *Informed consent forms distributed to participants, signatures secured, assurance of privacy/confidentiality/anonymity as appropriate, protection of the participant assurances reviewed, questions answered; note that the interview will be recorded and obtain permission for that, as well*

<u>Treatment of data:</u> *Researcher indicates how data will be managed, secured, and disposed of after a specific time period*

<u>Other questions or concerns?</u> *Other issues are discussed prior to beginning the interview session*

Opening the Interview Session

Introductory questions: Use these questions to introduce your topic and to establish a rapport with your participant.

Q1: Introductory question

> *Tell me briefly about your history with and current role within this department?*

Q2: Introductory question

> *In a few words or a phrase, how would you describe the department's primary mission and purpose?*

Key Interview Questions

The central portion of the interview consists of questions directly related to your research question and the elements of your topic that you wish

(Continued)

(Continued)

to explore. Remember to structure your questions from the broad to the specific in order to help your participant ease into the questioning route.

Q3. Content: *Can you outline the departmental operations and duties in a typical week from your perspective as a faculty member?*

Probes:

Q4. Content: *How would you describe the strengths, challenges, and opportunities for your department?*

Probes:

Q5. Content: *What is the nature of your relationships with your colleagues? Students? Program leadership? The department's position as an academic unit in the larger university?*

Q6. Content: *How do you perceive the culture of this department and its group self-consciousness? What are the unique features of the departmental culture? Traditions? Norms? Rules for behavior? How does that culture conform to or depart from the dominant university culture?*

Probes:

Concluding the Interview

Transition to the end of your interview session with one or two questions that allow the participant a chance to debrief or communicate any final thoughts, clarification, or comments that still need to be shared. A single open-ended question, posed by the researcher, is the best way to capture these final sentiments or thoughts.

Q7. Concluding question: *If you were to read a headline about your department in a major newspaper five years from now, what would you want it to say?*

Researcher script: *To obtain your final thoughts, is there anything else you would like to tell me or share with me regarding today's topic?*

Thank You and Follow-Up Reminder

Researcher Script: *Thank you for your time and your insights on (insert topic). I will follow-up with you in a few days to (choose one or more of the following) (1) ask you to complete a reflective questionnaire, (2) complete a member-checking exercise to verify my notes of our session, or (3) you a few questions for clarification.*

Appendix C Exemplar
Focus Group Moderator's Guide: Single Purpose

Information About the Focus Group

PARTICIPANTS (GENERAL): **_Student group-mixed_**
MODERATOR: **_Dr. X._** GROUP: **_Student Group #1_**
DATE: _____ TIME: _____ PLACE: _____

Introduction, Process, Consent

- Introduce yourself.
- Review the study's purpose, how long you expect the focus group to take, and your plans for using the results.
- Note that the interview will be audio-recorded and that you will keep their identities confidential.
- Distribute any profile survey questionnaires at this time, as appropriate to your study.

Ground Rules

Ground rules and group norms are always established at the beginning of a focus group session to ensure mutual respect, consideration, and a supportive atmosphere for the discussion:

- All group members have a right to their viewpoints and opinions.
- All group members have a right to speak without being interrupted or disrespected by other group members.
- Group members will avoid dominating the conversation and will allow time for others to speak.
- The moderator has the right to guide the timing and flow of the session topics but will allow the group to determine the importance and focus of the conversation, as appropriate.

(Continued)

(Continued)

- Identities of group members will remain confidential; first names only will be used for name tags and in reference to one another during the session.

Questioning Sequence

a. **Ice Breaker Question (60 seconds per participant)** Please share your first name, cohort year, and major.

b. **Introductory Question (90 seconds per participant)** If you were asked to create a slogan or catch phrase to describe this department, what would it be?

c. **Transition Question (1–2 sentences in description per participant)** How would you describe the defining features of this department? What are the important characteristics that distinguish it from other departments?

d. **Content Questions**

   Content #1: *How would you describe the ways the departmental faculty and staff support your academic goals? What are the strategies and practices they exhibit that contribute to your success? Or detract from your success?*

   Probes: Do you have specific examples or stories?

   Content #2: *Describe the ways in which the department creates or negates a sense of community for students.*

   Probes: Can you tell me more about that?

   Content #3: *What role does program and university leadership play in the department's growth and future success? What role should students play in the leadership of the department?*

   Probes: Do you have a particular story to share about that?

   Content #4: *How would you describe the departmental culture in terms of norms, traditions, distinctive traits and markings, rules for conduct and relationships? How does the departmental culture compare with your sense of the larger university culture?*

   Probes: Can you give more details?

Closing Question/Debriefing

5. What else would you like to tell me about?

Wrap Up and Thank You

- Thank you very much for your time today. I appreciated hearing your insights on this topic.
- If there is going to be a follow-up reflective process, please indicate that at this time.

Appendix D Exemplar
Observation Rubric for Formal or Informal Settings

Title of Project

TITLE OF STUDY: _____

DATE/TIME/DAY OF THE WEEK: _____

Number of participants: **55 (mixed)** Setting: **Department Town Meeting (Community)**

While observing the setting, the researcher will describe activity related to the following categories:

Relates to:	Individual behaviors	Group behaviors	Nonverbal cues	Conversation topics and threads
Participants Students Faculty PD Dean Community group	Students felt free to speak openly Faculty respected the student voice Dean was not present but PD facilitated with transparency	Groups merged naturally Some segregation between student cohorts Faculty dispersed among students evenly	Relaxed postures, considerable eye contact from PD to audience Students were not distracted with phones or each other too much Faculty were engaged and verbal	Program updates Problems with registration were reviewed Changes to advising process was introduced Student advisory council announced

Setting and use of space/ objects	Large room, comfortable space for people to sit at round tables	Collective sense of comfort with one another	Anxiety levels increased when talking about registration issues	
Types of ongoing activities	All attendees had an equal chance to participate Most attendees either presented or participated Some faculty dominated some parts of the conversation			
Demographic details	Just about 100% attendance; good sign that support is genuine for program/ department plans			
Researcher reflections	Considerable corroboration between positive feelings about department, distinctiveness of group culture			

Appendix E Exemplar
Combined Document/Artifact Rubric

Title of Project

Document/artifact	Location/source and author/creator	Original purpose of item	Date created	Consistent with findings	Divergence from findings
Doc-art 1 *Program Catalog, website, materials*	PD, faculty	Marketing, recruiting		Strongly consistent; few exceptions with spoken and written messages	Some catalog and program descriptions are not accurate, depart from original mission or emphasize something that stakeholders do not believe is possible/true
Doc-art 2 *Internal communications between PD and faculty, students*	PD, faculty, students	Community-building, sharing news and information		Strongly consistent	

Doc-art 3 *Program updates to university and external groups*	PD, Dean, university marketing and communications department	Outreach, branding, relationship-building	Consistent, aspirational in many ways though	Messages are more hopeful than real, but heading in the same direction as department's vision for the future
Doc-**art** 4 *Classroom spaces and meeting spaces*		Instruction, meeting, gathering	Classrooms were not conducive to goals for instruction, learning outcomes, or community-building—they were the only spaces available when program was created, will need to address for future	Significant divergence from the hopes, goals of faculty and students for instructional spaces and facilities
Doc-**art** 5 *Program offices, lounge, reception area*		Administration, welcoming, community	Administrative and welcoming spaces are more consistent with department mission	

Appendices 187

Appendix F Exemplar
Single Question Reflective Questionnaire

Title of Project

Thank you, again, for your willingness to meet with me regarding the formation and growth of Department X at XX University. I would like to ask you one final question to follow up on our discussion.

QUESTION:

Now that you have had a chance to reflect further on our discussion regarding the ways in which Department X has grown and flourished over the past five years, is there anything else you wish to share regarding your perceptions of how the department can continue on this successful path? In what ways can you, personally, contribute to the department's future?

Please feel free to share any thoughts, details, examples, or experiences that you believe will add to our discussion on this topic. Your insights and your perceptions are important and will add richness to this study's findings.

I would appreciate your response within the next week, by (insert date). Please feel free to contact me if you have any questions or wish to talk further.

Thank you,
Signature/researcher's contact information

_____

_____

Appendix G Exemplar
Data Collection Plan

Data Sources	Date/Time	IRB	Interviews	Focus Groups	Documents/Artifacts	Observations	Quest
Pilot interviews							
Elite	xx/xx	Y	X				
Interviews-Faculty	xx/xx–xx/xx	Y	X				
Focus Groups-Students	xx/xx	Y		X			
Observation-Community Mtg	xx/xx	Y				X	
Documents/Artifacts-Program	xx/xx	Y	X		X		
Reflective Questionnaires	xx/xx	Y	X				X
Phases							
Phase I, Elite Interview	xx/xx	Y	X				
Phase II, Interviews-Faculty							

(*Continued*)

(Continued)

Phase III, Focus Groups-Students						
Group 1	xx/xx-xx/xx	Y	X			
Group 2		Y				
Group 3		Y				
Phase IV, Observation	xx/xx	Y			X	
Phase V, Document/Artifact Analysis		Y				
Phase VI, Reflective Questionnaires		Y				

NOTE: This matrix is only partially completed but illustrates how it can be used to track data collection and the progress of the study.

Recommended Qualitative Research Websites

The following websites are recommended as comprehensive resources designed for qualitative researchers. While many websites relate to *specific* aspects of qualitative research, the following websites serve as valuable compendiums and are useful as starting points for anyone wishing to learn more about qualitative research methods and applications.

International Institute for Qualitative Methodology (2018)

https://www.ualberta.ca/international-institute-for-qualitative-methodology

The International Institute for Qualitative Methodology (IIQM) is an interdisciplinary institute based at the University of Alberta, in Edmonton, Alberta, Canada, serving qualitative researchers around the world. IIQM was founded in 1998, with the primary goal of facilitating the development of qualitative research methods across a wide variety of academic disciplines. The website outlines a list of current conferences, workshops, webinars, and resources, as well as information about Member Scholar and distinguished/visiting scholar programs.

The Qualitative Report, Nova University

https://tqr.nova.edu/websites/

The Qualitative Report first started as a weekly page to better leverage resources and opportunities for qualitative researchers. The weekly community page now features select articles from the upcoming monthly journal publication, conference news, featured blogs, and other useful resources for new and experienced qualitative researchers.

Qualitative Research Consultants Association

https://www.qrca.org/page/about-qrca

QRCA is a not-for-profit association of consultants involved in the design and implementation of qualitative research applications. This global association of professionals is dedicated to promoting excellence in the field of qualitative research by pooling experience and expertise to create a base of shared knowledge. A list of events, resources, and links is available and regularly updated.

References

Agar, M. H. (1980). *The professional stranger: An informal introduction to ethnography.* San Diego, CA: Academic Press.

Agee, J. (2009). Developing qualitative research questions: A reflective process. *International Journal of Qualitative Studies in Education, 431*–447.

Ahern, K. (1999). Ten tips for reflexive bracketing. *Qualitative Health Research, 9*, 407–411.

Alaszewski, A. (2006). *Using diaries for social research.* London, England: Sage.

Altheide, D. L., & Johnson, J. M. (1994). Criteria for assessing interpretive validity in qualitative research. In N. K. Denzin and Y. S. Lincoln (Eds.), *Handbook of qualitative research* (pp. 485–499). Thousand Oaks, CA: Sage.

Alvesson, M. (2002). *Understanding organizational culture.* London, UK: Sage.

Alvesson, M. (2012). *Interpreting interviews.* Thousand Oaks, CA: Sage.

Alvesson, M., & Skoldberg, K. (2010). *Reflexive methodology: New vistas for qualitative research* (2nd ed.). Thousand Oaks, CA: Sage.

Angen, M. J. (2000). Evaluating interpretive inquiry: Reviewing the validity debate and opening the dialogue. *Qualitative Health Research, 10*(3), 378–395.

Angers, J., & Machtmes, K. L. (2005). An ethnographic-case study of beliefs, context factors, and practices of teachers integrating technology. *The Qualitative Report, 10*(4), 771–794.

Angrosino, M. (2014). *Doing ethnographic and observational research.* Thousand Oaks, CA: Sage.

Angrosino, M. V., & de Perez, K. A. (2000). Rethinking observation: From method to context. In N. K. Denzin & Y. S. Lincoln (Eds.), *Handbook of qualitative research* (2nd ed.), (pp. 673–702). Thousand Oaks, CA: Sage.

Arp, R. (2004). Husserl and the penetrability of the transcendental and mundane spheres. *Human Studies, 27*(3), 221–239.

Atkinson, R. (2016). *The life story interview.* Thousand Oaks, CA: Sage.

Bailey, C. A. (2018). *A guide to qualitative field research* (3rd ed.). Thousand Oaks, CA: Sage.

Barbour, R. S. (2007). *Doing focus groups.* Thousand Oaks, CA: Sage.

Barbour, R. S., & Kitzinger, J. (1999). *Developing focus group research: Politics, theory and practice.* London, UK: Sage.

Bau, V. (2016). A narrative approach in evaluation: "Narratives of Change" method. *Qualitative Research Journal, 16*(4), 374–387.

Berger, R. (2015). Now I see it, now I don't: Researcher's position and reflexivity in qualitative research. *Qualitative Research, 15*(2), 219–234.

Bernard, H. R. (1988). *Research methods in cultural anthropology.* Thousand Oaks, CA: Sage.

Bernard, H. R. (1994). *Research methods in anthropology: Qualitative and quantitative approaches.* Thousand Oaks, CA: Sage.

Bernard, H. R. (2013). *Social research methods: Qualitative and quantitative approaches* (2nd ed.). Thousand Oaks, CA: Sage.

Bernard, H. R., Wutich, A., & Ryan, G. W. (2017). *Analyzing qualitative data: Systematic approaches* (2nd ed.). Thousand Oaks, CA: Sage.

Billups, F. D. (Spring, 2012). Qualitative data analysis: An overview for beginning qualitative researchers. *The NERA Researcher, 50*(1), 8–10.

Billups, F. D. (Summer/Fall, 2013). Focus group research: What makes a focus group . . . A focus group? *The NERA Researcher, 51,* 10–11.

Billups, F. D. (Fall, 2014). Trustworthiness and the quest for rigor in qualitative research. *The NERA Researcher, 52,* 10–12.

Billups, F. D. (Fall, 2016). Applying Kruger and Casey's classic approach to focus group data analysis. *The NERA Researcher,* 10–11.

Birks, M., & Mills, J. (2011). *Grounded theory: A practical guide.* London, UK: Sage.

Bischoping, K, & Gazso, A. (2015*). Analyzing talk in the social sciences: Narrative, conversation and discourse strategies.* Thousand Oaks, CA: Sage.

Boeije, H. (2010). *Analysis in qualitative research.* London, UK: Sage.

Bogdan, R., & Bilken, S. K. (2003*). Qualitative research for education: An introduction to theory and methods.* Boston, MA: Allyn and Bacon.

Bowen, G. A. (2009). Document analysis as a qualitative research method. *Qualitative Research Journal, 9*(2), 27–40.

Boyatzis, R. E. (1998). *Transforming qualitative information: Thematic analysis and code development.* Thousand Oaks, CA: Sage.

Brinkmann, S., & Kvale, S. (2014). *Learning the craft of qualitative research interviewing* (3rd ed.). Thousand Oaks, CA: Sage.

Britten, N., Jones, R., Murphy, E., & Stacy, R. (1995). Qualitative research methods in general practice and primary care. *Family Practice, 12*(1), 104–114.

Brock, B. L. (1985). Epistemology and ontology in Kenneth Burke's Dramatism. *Communication Quarterly, 33,* 94–104.

Brody, J. L., Gluck, J. P., & Aragon, A. S. (2000). Participants' understanding of the process of psychological research: Debriefing. *Ethics and Behavior, 10*(1), 13–25.

Brundage, A. (2017). *Going to the sources: A guide to historical research and writing* (6th ed.). Hoboken, NJ: Wiley Blackwell.

Bryant, A., & Charmaz, K. (2007). Grounded theory in historical perspective: An epistemological account. In A. Bryant & K. Charmaz (Eds.), *The Sage handbook of grounded theory* (pp. 31–57). Thousand Oaks, CA: Sage.

Burke, K. (1969). *Language as symbolic action: Essays on life, literature, and method*. Berkeley: University of California Press.

Catalani, C., & Minkler, M. (2010). Photovoice: A review of the literature in health and public health. *Health Education & Behavior, 37*(3), 424–451.

Charmaz, K. (2014). *Constructing grounded theory* (2nd ed.). Thousand Oaks, CA: Sage.

Clandinin, D. J. (2013). *Engaging in narrative inquiry*. New York, NY: Routledge.

Clandinin, D. J., & Connelly, F. M. (2000). *Narrative inquiry: Experience and story in qualitative research*. San Francisco, CA: Jossey-Bass.

Clark, A., & Emmel, N. (2009). Connected lives: Methodological challenges for researching networks, neighbourhoods and communities. *Qualitative Researcher, 11*, 9–11.

Clarke, A. E., Friese, C., & Washburn, R. S. (2017). *Situational analysis: Grounded theory after the interpretive turn* (2nd ed.). Thousand Oaks, CA: Sage.

Cohen, D., & Crabtree, B. (2006). *Qualitative research guidelines project*. Retrieved from http://www.qualres.org/HomeEval-3664.html.

Colaizzi, P. F. (1978). Psychological research as the phenomenologist views it. In R. S. Valle & M. King (Eds.), *Existential-phenomenological alternatives for psychology*. Oxford, UK: Oxford University Press.

Corbin, J., & Morse, J. M. (2003). The unstructured interactive interview: Issues of reciprocity and risks when dealing with sensitive topics. *Qualitative Inquiry, 9*, 335–354.

Corbin, J., & Strauss, A. (2015). *Basics of qualitative research: Techniques and procedures for developing grounded theory* (3rd ed.). Thousand Oaks, CA: Sage.

Cortazzi, M. (1993). *Narrative analysis*. London, UK: Falmer Press.

Crabtree, B. F., & Miller, W. L. (2015). *Doing qualitative research*. Thousand Oaks, CA: Sage.

Creswell, J. W. (2013). *Qualitative inquiry & research design: Choosing among the five approaches* (3rd ed.). Thousand Oaks, CA: Sage.

Creswell, J. W., & Poth, C. N. (2018). *Qualitative inquiry & research design: Choosing among the five approaches* (4th ed.). Thousand Oaks, CA: Sage.

Czarniawska, B. (1997). *Qualitative research methods: A narrative approach to organization studies*. Thousand Oaks, CA: Sage.

Daiute, C. (2013). *Narrative inquiry: A dynamic approach*. Thousand Oaks, CA: Sage.

de Laine, M. (2000). *Fieldwork, participation, and practice: Ethics and dilemmas in qualitative research.* London, UK: Sage.

Denzin, N. K. (1970). *The research act: A theoretical introduction to sociological methods.* New York, NY: Aldine.

Denzin, N. K. (2001). The reflexive interview and a performative social science. *Qualitative Research, 1,* 23–46.

Denzin, N. K., & Lincoln, Y. S. (Eds.). (2003). *Turning points in qualitative research.* Lanham, MD: AltaMira Press.

Denzin, N. K., & Lincoln, Y. S. (2011). Introduction: The discipline and practice of qualitative research. In N. K. Denzin & Y. S. Lincoln (Eds.), *The Sage handbook of qualitative research* (4th ed.), (pp. 1–9). Thousand Oaks, CA: Sage.

Denzin, N. K., & Lincoln, Y. S. (2013). *Strategies of qualitative inquiry.* Thousand Oaks, CA: Sage.

DeWalt, K. M., & DeWalt, B. R. (2002). *Participant observation: A guide for fieldworkers.* Plymouth, UK: AltaMira Press.

DeWalt, K. M., & DeWalt, B. R. (2011). *Participant observation: A guide for fieldworkers* (2nd ed.). Plymouth, UK: AltaMira Press.

DiCiccio-Bloom, B., & Crabtree, B. F. (2006). The qualitative research interview. *Medical Education, 40,* 314–321.

Doody, O., & Noonan, M. (2013). Preparing and conducting interviews to collect data. *Nurse Researcher, 20*(5), 28–32.

Elder, N. C., & Miller, W. L. (1995). Reading and evaluating qualitative research studies. *The Journal of Family Practice, 41*(3), 279–285.

Eppley, K. (2006). Defying insider-outsider categorization: One researcher's fluid and complicated positioning on the insider-outsider continuum. *Forum: Qualitative Social Research, 7*(3), 1–8.

Epstein, I., Stevens, B., McKeever, P., & Baruchel, S. (2006). Photo elicitation interview (PEI): Using photos to elicit children's perspectives. *International Journal of Qualitative Methods, 5*(3), 1–11.

Fairlough, N. (2010). *Critical discourse analysis: The critical study of language* (2nd ed.). New York, NY: Routledge

Fairlough, N. (2015). *Language and power* (3rd ed.). New York, NY: Routledge.

Fern, E. F. (2001). *Advanced focus group research.* Thousand Oaks, CA: Sage.

Fetterman, D. (1998). *Ethnography: Step-by-step* (2nd ed.). Applied Social Research Methods Series, Vol. 17. Newbury Park, CA: Sage.

Fetterman, D. (2010). *Ethnography: Step-by-step* (3rd ed.). Applied Social Research Methods Series, Vol. 17. Thousand Oaks, CA: Sage.

Finlay, L. (2008). Reflecting on "reflective practice." *Practice-based Professional Learning Centre,* Paper 52, 1–24.

Flick, U. (2009). *Introduction to qualitative research* (4th ed.). Thousand Oaks, CA: Sage.

Frey, J. H., & Oishi, S. M. (1995). *How to conduct interviews by telephone and in person*. Thousand Oaks, CA: Sage.

Gall, M. D., Gall, J. P., & Borg, W. R. (2006). *Educational research* (8th ed.). Boston, MA: Pearson.

Gee, J. P. (1991). A linguistic approach to narrative. *Journal of Narrative and Life History/Narrative Inquiry, 1*, 15–39.

Gee, J. P. (2014a). *An introduction to discourse analysis: Theory and method* (4th ed.). New York, NY: Routledge.

Gee, J. P. (2014b). *How to do discourse analysis: A toolkit* (2nd ed.). New York, NY: Routledge.

Geertz, C. (1973). *The interpretation of cultures*. New York, NY: Basic Books.

Gill, P., Stewart, K., Treasure, E., & Chadwick, B. (2008). Methods of data collection in qualitative research: Interviews and focus groups. *British Dental Journal, 204*, 291–295.

Gioia, D. A., Corley, K. G., & Hamilton, A. L. (2012). Seeking qualitative rigor in inductive research: Notes on the Gioia methodology. *Organizational Research Methods, 16*(1), 15–31.

Giorgi, A. (1985). *Phenomenology and psychological research*. Pittsburgh, PA: Duquesne University Press.

Giorgi, A. (1994). A phenomenological perspective on certain qualitative research methods. *Journal of Phenomenological Psychology, 25*, 190–220.

Glaser, B. G. (Ed.). (1993). *Examples of grounded theory: A reader*. Mill Valley, CA: Sociology Press.

Glaser, B. G. (2000). The future of grounded theory. *Grounded Theory Review, 1*, 1–18.

Glaser, B. G., & Strauss, A. (1967). *The discovery of grounded theory: Strategies for qualitative research*. Chicago, IL: Aldine.

Gold, R. L. (1958). Roles in sociological field observations. *Social Forces, 36*, 217–223.

Goodyear, V. A., Casey, A., & Quennerstedt, M. (2018). Social media as a tool for generating sustained and in-depth insights into sport and exercise practitioners' ongoing practices. *Qualitative Research in Sport, Exercise and Health, 10*(1), 1–16.

Grbich, C. (2012). *Qualitative data analysis: An introduction*. Thousand Oaks, CA: Sage.

Guba, E. G. (Ed.). (1990). *The paradigm dialog*. Newbury Park, CA: Sage.

Gubrium, J. F., & Holstein, J. A. (Eds.). (2003). *Postmodern interviewing*. Thousand Oaks, CA: Sage.

Gubrium, J. F., & Holstein, J. A. (2013). *Analyzing narrative reality*. Thousand Oaks, CA: Sage.

Halquist, D., & Musanti, S. I. (2010). Critical incidents and reflection: Turning points that challenge the researcher and create opportunities for knowing. *International Journal for Qualitative Studies in Education*, 449–461.

Hamel, J., Dufour, S., & Fortin, D. (1993). *Qualitative research methods: Case study methods*. Thousand Oaks, CA: Sage.

Hammersley, M. (1990). *Reading ethnographic research: A critical guide*. New York, NY: Longman.

Hammersley, M., & Atkinson, P. (2007). *Ethnography: Principles in practice*. London, UK: Routledge.

Harris, J. (2016). Utilizing the walking interview to explore campus climate for students of color. *Journal of Student Affairs Research and Practice, 53*(4), 365–377.

Heritage, J. (1988). Current developments in conversation analysis. In D. Roger & P. Bull (Eds.), *Conversation: An interdisciplinary approach* (pp. 21–47). Clevedon, UK: Multilingual Matters.

Heritage, J. (2004). Conversation analysis and institutional talk: Analyzing data. In D. Silverman (Ed.), *Qualitative research: Issues of theory, method, and practice* (3rd ed.), (pp. 222–245). London, UK: Sage.

Hodder, I. (2003). The interpretation of documents and material culture. In N. K. Denzin & Y. S. Lincoln (Eds.), *Collecting and interpreting qualitative materials* (2nd ed.), (pp. 155–175). Thousand Oaks, CA: Sage.

Hughes, J., & Goodwin, J. (2014). *Documentary & archival research*. Thousand Oaks, CA: Sage.

Husserl, E. (1970). *The crisis of European sciences and transcendental phenomenology* (D. Carr, Trans.). Evanston, IL: Northwestern University Press.

Hutchby, I., & Wooffitt, R. (2008). *Conversation analysis: Principles, practices, and applications*. Cambridge, UK: Polity Press.

James, N. (2013). The use of e-mail interviewing as a qualitative method of inquiry in educational research. *British Educational Research Journal, 33*(6), 963–976.

James, N. (2016). Using e-mail interviews in qualitative educational research: Creating space to think and time to talk. *International Journal of Qualitative Studies in Education, 29*(2), 150–163.

James, N. (2017). You've got mail . . . ! Using e-mail interviews to gather academics' narratives of their working lives. *International Journal of Research & Method in Education, 40*(1), 6–18.

Johnstone, B. (2008). *Discourse analysis* (2nd ed.). Malden, MA: Blackwell.

Josselson, R., & Lieblich, A. (2015). *Making meaning of narratives*. Thousand Oaks, CA: Sage.

Kahn, R. L., & Cannell, C. F. (1957). *The dynamics of interviewing; theory, technique, and cases*. Oxford, UK: John Wiley & Sons.

Kawulich, B. B. (2005). Participant observation as a data collection method. *Forum: Qualitative Social Research, 6*(2), 1–8.

Keller, R. (2013). *Doing discourse research: An introduction for social scientists*. Thousand Oaks, CA: Sage.

King, N., & Horrocks, C. (2011). *Interviews in qualitative research*. Thousand Oaks, CA: Sage.

Kitzinger, J., & Barbour, R. S. (1999). Introduction: The challenge and promise of focus groups. In R. S. Barbour &

J. Kitzinger (Eds.), *Developing focus group research: Politics, theory and practice* (pp. 36–46). London, UK: Sage.

Krefting, L. (1991). Rigor in qualitative research: The assessment of trustworthiness. *The American Journal of Occupational Therapy, 45,* 214–222.

Krippendorff, K., & Bock, M. A. (2009). *The content analysis reader.* Thousand Oaks, CA: Sage.

Krueger, R. A., & Casey, M. A. (2015). *Focus groups: A practical guide for applied research* (5th ed). Thousand Oaks, CA: Sage.

Kuckartz, U. (2014). *Qualitative text analysis: A guide to methods, practice and using software.* Thousand Oaks, CA: Sage.

Kutsche, P. (1998). *Field ethnography: A manual for doing cultural anthropology.* Upper Saddle River, NJ: Prentice-Hall.

Kvale, S. (2012). *Doing interviews.* Thousand Oaks, CA: Sage.

Kvale, S., & Brinkmann, S. (2009). *Interviews: Learning the craft of qualitative research interviewing* (2nd ed.). Thousand Oaks, CA: Sage.

Kvale, S., & Brinkmann, S. (2014). *Interviews: Learning the craft of qualitative research interviewing* (3rd ed.). Thousand Oaks, CA: Sage.

Labov, W. (2006). Narrative preconstruction. *Narrative Inquiry, 16,* 37–45.

Lange, M. (2012). *Comparative-historical methods.* Thousand oaks, CA: Sage.

Lester, J. N., & O'Reilly, M. (2018). Applied conversation analysis: Social interaction in institutional settings. Thousand Oaks, CA: Sage.

Liamputtong, P. (2011). *Focus group methodology: Principles and practice.* Thousand Oaks, CA: Sage.

Lincoln, Y. S., & Denzin, N. K. (Eds.). (2008). *Collecting and interpreting qualitative materials* (3rd ed.). Thousand Oaks, CA: Sage.

Lincoln, Y. S., & Guba, E. G. (1985). *Naturalistic inquiry.* Newbury Park, CA: Sage.

Livholts, M., & Tamboukou, M. (2015). *Discourse and narrative methods: Theoretical departures, analytical strategies and situated writings.* Thousand Oaks, CA: Sage.

Lofland, J., & Lofland, L. (1995). *Analyzing social settings: A guide to qualitative observation and analysis.* Belmont, CA: Wadsworth.

Lypka, A. E. (2017). Demystifying the analysis process of talk data: A review of analyzing talk in the social sciences. Narrative, conversation, & discourse strategies. *The Qualitative Report, 22*(3), 868–872.

Madison, D. S. (2012). *Critical ethnography: Method, ethics, and performance* (2nd ed.). Thousand Oaks, CA: Sage.

Malterud, K. (2001). Qualitative research: Standards, challenges and guidelines. *The Lancet, 358,* 483–488.

Marshall, C., & Rossman, G. B. (1989). *Designing qualitative research.* Thousand Oaks, CA: Sage.

Marshall, C., & Rossman, G. B. (2011). *Designing qualitative research* (5th ed.). Thousand Oaks, CA: Sage.

Maxwell, J. A. (2005). *Qualitative research design: An interactive approach* (2nd ed.). Thousand Oaks, CA: Sage.

Maxwell, J. A. (2013). *Qualitative research design: An interactive approach* (3rd ed.). Thousand Oaks, CA: Sage.

McDowell, S. (2013). *Historical research: A guide for writers of dissertations, theses, articles, and books*. New York, NY: Routledge.

Merriam, S. B. (2002). *Qualitative research in practice: Examples for discussion and analysis*. San Francisco, CA: Jossey-Bass.

Merriam, S. B., & Tisdell, E. J. (2015). *Qualitative research: A guide to design and implementation* (4th ed.). San Francisco, CA: Jossey-Bass.

Miles, M. B., & Huberman, A. M. (1994). *Qualitative data analysis: A sourcebook of new methods* (2nd ed.). Thousand Oaks, CA: Sage.

Miles, M. B., Huberman, A. M., & Saldana, J. (2014). *Qualitative data analysis: A methods sourcebook* (3rd ed.). Thousand Oaks, CA: Sage.

Miller, R. L. (2015). *Researching life stories and family histories*. Thousand Oaks, CA: Sage.

Minichiello, V., Aroni, R., & Hays, T. (2008). *In-depth interviewing: Principles, techniques, analysis*. Frenchs Forest, Sydney: Pearson Education Australia.

Morgan, D. L. (1997). *Focus groups as qualitative research* (2nd ed.). Thousand Oaks, CA: Sage.

Morgan, D. L. (2016). *Essentials of dyadic interviewing*. Walnut Creek, CA: Left Coast Press.

Morgan, D. L., & Krueger, R. A. (Eds.). (1998). *The focus group kit*. Thousand Oaks, CA: Sage.

Morgan, D. L, Ataie, J., Carder, P., & Hoffman, K. (2013). Introducing dyadic interviews as a method for collecting qualitative data. *Qualitative Health Research, 23*(9), 1276–1284.

Morris, A. (2015). *A practical introduction to in-depth interviewing*. Thousand Oaks, CA: Sage.

Morse, J. M., Barrett, M., Mayan, M., Olson, K., & Spiers, J. (2002). Verification strategies for establishing reliability and validity in qualitative research. *International Journal of Qualitative Methods, 1*(2), 1–19.

Moustakas, C. (1994). *Phenomenological research methods*. Thousand Oaks, CA: Sage.

O'Cathain, A., & Thomas, K. J. (2004). Any other comments? Open questions on questionnaires—A bane or a bonus to research? *Medical Research Methodology, 4*(25), 1–7.

O'Leary, Z. (2014). *The essential guide to doing your research project* (2nd ed.). Thousand Oaks, CA: Sage.

Ortlipp, M. (2008). Keeping and using reflective journals in the qualitative research process. *The Qualitative Report, 13*(4), 695–705.

Patton, M. Q. (1999). Enhancing the quality and credibility of qualitative analysis. *HSR: Health Services Research, 34*(5), 1189–1208.

Patton, M. Q. (2002). *Qualitative research & evaluation methods: Integrating theory and practice* (3rd ed.). Thousand Oaks, CA: Sage.

Patton, M. Q. (2015). *Qualitative research & evaluation methods: Integrating theory and practice* (4th ed.). Thousand Oaks, CA: Sage.

Pennebaker, J. W., & Seagal, J. D. (1999). Forming a story: The health benefits of narrative. *Journal of Clinical Psychology, 55*(10), 1243–1254.

Phillips, N., & Hardy, C. (2015). *Discourse analysis: Investigating processes of social construction.* Los Angeles, CA: Sage.

Pinnegar, S., & Daynes, J. G. (2007). Locating narrative inquiry historically: Thematic in the turn to narrative. In J. Clandinin (Ed.), *Handbook of narrative inquiry: Mapping a methodology* (pp. 3–34). Thousand Oaks, CA: Sage.

Pivčević, E. (2014). *Husserl and phenomenology.* London, UK: Routledge.

Pizam, A. (1994). Planning a tourism research investigation. In J. R. Ritchie & C. R. Goeldner (Eds.), *Travel, tourism, and hospitality research: A handbook for managers and researchers.* New York, NY: John Wiley & Sons.

Polit, D. F., & Beck, C. T. (2017). *Nursing research: Generating and assessing evidence for nursing practice* (10th ed.). Philadelphia, PA: Lippincott, Williams, & Wilkins.

Potter, J., & Wetherell, M. (2005). *Discourse and social psychology: Beyond attitudes and behavior.* London, UK: Sage.

Prior, L. (2014). *Using documents in social research.* Thousand Oaks, CA: Sage.

Punch, M. (1986). *The politics and ethics of fieldwork.* Qualitative Research Methods Series, No. 3. London, UK: Sage.

Rapley, T. (2011). *Doing conversation, discourse and document analysis.* Thousand Oaks, CA: Sage.

Reid, K., Flowers, P., & Larkin, M. (2005). Exploring lived experience. *The Psychologist, 18*(1), 20–23.

Riessman, C. K. (1993). *Narrative analysis.* Thousand Oaks, CA: Sage.

Riessman, C. K. (1996). *Qualitative research methods: Narrative analysis.* Thousand Oaks, CA: Sage.

Riessman, C. K. (2008). *Narrative methods for the human sciences.* Thousand Oaks, CA: Sage.

Rogers, R. (2004). An introduction to critical discourse analysis in education. In *An introduction to critical discourse analysis in education* (pp. 31–48). New York, NY: Routledge.

Roulston, K. (2012). *Reflective interviewing: A guide to theory and practice* (2nd ed.). Thousand Oaks, CA: Sage.

Roulston, K. (2013). Interactional problems in research interviews. *Qualitative Research, 14*(3), 277–293.

Rubin, H. J., & Rubin, I. S. (2013). *Qualitative interviewing: The art of hearing data* (3rd ed.). Los Angeles, CA: Sage.

Ryle, G. (1949). *The concept of mind.* London, UK: Hutchinson.

Sacks, H., Schegloff, E. A., & Jefferson, G. (1974). A simplest systematics for the organization of turn taking in conversation. *Language, 50*(4), 696–735.

Saldana, J. (2015). *Thinking qualitatively: Methods of mind.* Thousand Oaks, CA: Sage.

Schwandt, T. (2015). *The Sage dictionary of qualitative inquiry* (4th ed.). Thousand Oaks, CA: Sage.

Schwartzman, H. B. (1992). *Qualitative research methods: Ethnography in organizations.* Thousand Oaks, CA: Sage.

Seidman, I. (2013). *Interviewing as qualitative research: A guide for researchers in education and the social sciences* (4th ed.). New York, NY: Teachers College Press.

Sewell, W. H., Jr. (2005). *Logics of history: Social theory and social transformation.* Chicago, IL: University of Chicago Press.

Silverman, D. (2008). *Doing qualitative research: A comprehensive guide.* Thousand Oaks, CA: Sage.

Silverman, D. (2013). What counts as qualitative research? Some cautionary comments. *Qualitative Sociology Review, 9*, 48–55.

Spradley, J. P. (1979). *The ethnographic interview.* Belmont, CA: Wadsworth/Cengage Learning.

Spradley, J. P. (2016, reissued). *Participant observation.* Long Grove, IL: Waveland Press.

Stake, R. E. (1995). *The art of case study research.* Thousand Oaks, CA: Sage.

Stebbins, R. A. (2001). *Qualitative research methods: Exploratory research in the social sciences.* Thousand Oaks, CA: Sage.

Stewart, A. (1998). *Qualitative research methods: The ethnographer's method.* Thousand Oaks, CA: Sage.

Stewart, D. W., & Shamdasani, P. N. (2014). *Focus groups: Theory and practice* (3rd ed.). Thousand Oaks, CA: Sage.

Strauss, A. L., & Corbin, J. M. (1998). *Basics of qualitative research: Techniques and procedures for developing grounded theory.* Thousand Oaks, CA: Sage.

Sullivan, P. (2013). *Qualitative data analysis using a dialogical approach.* Thousand Oaks, CA: Sage.

ten Have, P. (2007). *Doing conversation analysis: A practical guide.* Thousand Oaks, CA: Sage.

ten Have, P. (2012). *Doing conversation analysis: A practical guide* (2nd ed.). Thousand Oaks, CA: Sage.

Thomas, G. (2015). *How to do your case study* (2nd ed.). Thousand Oaks, CA: Sage.

Tinkler, P. (2013). *Using photographs in social and historical research.* Thousand Oaks, CA: Sage.

Trochim, W. M. K. (2006). *Research methods knowledge base.* New York, NY: Concept Systems Knowledge Base. Retrieved from http://www.socialresearchmethods.net/kb/index.php.

Tseliou, E. (2013). A critical methodological review of discourse and conversation analysis studies of family therapy. *Family Process, 52*(4), 653–672.

Van Dijk, T. A. (1997). *Discourse as social interaction.* Los Angeles, CA: Sage.

van Lier, L. (1988). *The classroom and the language learner: Ethnography and second language classroom research.* New York, NY: Longman.

van Manen, M. (2014). *Phenomenology of practice.* Walnut Creek, CA: Left Coast Press.

Van Maanen, J. (1978). Epilogue: On watching the watchers. In P. K., and J. Van Maanen (Eds.), *Policing: A view from the street* (pp. 309–349). Santa Monica, CA: Goodyear.

Van Maanen, J. (2011). *Tales of the field: On writing ethnography* (2nd ed.). Chicago, IL: University of Chicago Press.

Van Maanen, J., & Barley, S. (1985). Cultural organizations: Fragments of a theory. In P.J. Frost, L. F. Moore, M. R. Louis, C. C. Lundberg, & J. Martin (Eds.), *Organizational culture* (pp. 31–54). Newbury Park, CA: Sage.

Vaughn, S., Schumm, J. S., & Sinagub, J. (1996). *Focus group interviews in education and psychology.* Thousand Oaks, CA: Sage.

Wallendorf, M., & Belk, R. W. (1989). Assessing trustworthiness in naturalistic consumer research. In E. C. Hirschman, (Ed.), *SV—Interpretative consumer research* (pp. 69–84). Provo, UT: Association for Consumer Research Retrieved from http://acrwebsite.org/volumes/12177/volumes/sv07/SV-07.

Wang, X. (2013). The construction of researcher-researched relationships in school ethnography: Doing research, participating in the field and reflecting on ethical dilemmas. *International Journal of Qualitative Studies in Education, 26*(7), 763–779.

Wang, C., & Burris, M. (1997). Photovoice: Concept, methodology, and use or participatory needs assessment. *Health Education & Behavior, 24*(3), 369–387.

Weller, S. C., & Romney, A. K. (1988). *Qualitative research methods: Systematic data collection.* Thousand Oaks, CA: Sage.

Werner, D., & Schoeppfle, G. M. (1987). *Ethnographic analysis and data management* (Vol. 2). Thousand Oaks, CA: Sage.

Whittemore, R., Chase, S. K., & Mandle, C. L. (2001). Validity in qualitative research. *Qualitative Health Research, 11,* 117–132.

Wiggins, S. (2017). *Discursive psychology: Theory, method and applications.* Thousand Oaks, CA: Sage.

Wolcott, H. F. (2008). *Ethnography: A way of seeing.* Walnut Creek, CA: AltaMira Press.

Wolcott, H. F. (2009). *Writing up qualitative research* (3rd ed.). Thousand Oaks, CA: Sage.

Wood, L. A., & Kroger, R. O. (2015). *Doing discourse analysis: Methods for studying action in talk and text.* Thousand Oaks, CA: Sage.

Wooffitt, R. (2014). *Conversation analysis and discourse analysis: A comparative and critical introduction.* Thousand Oaks, CA: Sage.

Yussen, S. R., & Ozcan, N. M. (1996). The development of knowledge about narratives. *Issues in Education: Contributions from Educational Psychology, 2,* 1–68.

Index

Tables are indicated by t following the page number.

Accuracy, 28t
Agar, M. H., 61
Alvesson, M., 36, 155
Ambiguity, 173
Analyst triangulation, 29
Angrosino, M., 134, 145
A priori coding, 83
Artifact and document rubrics, 12t, 147–151, 186–187
Artifacts and documents, 143–147, 144t
Assumptions, 25
ATALIS.ti, 164
Attention probes, 47
Audio recordings, 81–82, 136
Audits, 30
Audit trails, 31
Authenticity, 28t, 31
Axiological worldview, 2–3

Bailey, C. A., 164
Barrett, M., 31
Berger, R., 155
Bernard, H. R., 43, 134
Bias, 24–25, 39–40, 146. *See also* Reflexivity
Billups, F. D., 132
Boundary violations, 32
Bowen, G. A., 143, 145, 146, 147, 151
Bracketing, 25
Brainstorming focus groups, 100t, 111t, 120–122
Brinkmann, S., 36
Burke, K., 69–70

Case studies
  overview, 6–7, 8t
  data collection and, 21t, 170t
  guiding questions and, 18t
  research questions and, 19t
Case study example
  data collection plan for, 189–190
  document and artifact rubric for, 186–187
  focus group moderator guide for, 181–183
  observation rubric for, 184–185
  reflective questionnaire for, 188
  research questions for, 175–176
  semistructured interview protocol for, 179–180
  unstructured interview protocol for, 177–178
Casey, M. A., 96, 128–129
Clandinin, D. J., 68
Clarification probes, 47
Classic Approach strategy, 129–130
Clustering, 83–84
Coding of data, 83–84
Concluding questions, 103–104
Confirmability, 28t, 30–31
Content questions, 103
Continuation probes, 47
Contrast questions, 64
Conversational/discourse analysis, 87–94
Conversational interviews. *See* Unstructured interviews
Conversation/discourse logs, xxv, 12t, 89–94
Crabtree, B. F., 42
Creativity, 172–173
Credibility, 28t, 29–30
Creswell, J. W., 23, 68, 154, 155, 166
Critical ethnography, 62
Critical incidents framework, 130
Cultural analysis, 61, 89
Cultural/physical objects, 144t, 146–147, 149. *See also* Documents and artifacts
Cultural questions, 64

Data analysis
  conversational/discourse analysis and, 94
  documents and artifacts, 151
  focus groups and, 128–130
  interview protocols and, 82–85
  observation rubrics and, 141–142
  reflective questionnaires and, 164–165

Data cleaning, 83
Data collection plans, 170–172, 171t
Data reduction, 83–84
Data sources, 3, 10, 10–11t, 26–27
Data triangulation, 29
Debriefing questions, 103–104
de Laine, M., 32
Denzin, N. K., 1, 144, 147
Dependability, 28t, 30
de Perez, K. A., 134, 145
Depth interviews, 42
Descriptive/interpretive research design, 5, 7t, 17t, 19t, 20t, 169t
Descriptive questions, 64
DeWalt, B. R., 133
DeWalt, K. M., 133
Diaries, xxv, 157, 157t, 162–163, 164
DiCiccio-Bloom, B., 42
Discourse/conversation analysis, 87–94
Discourse/conversation logs, xxv, 12t, 89–94
Document and artifact rubrics, 12t, 147–151, 186–187
Documents and artifacts, 143–147, 144t
Double-layered focus groups, 100t
Dual moderators focus groups, 100t, 110t, 116–117
Dueling moderators focus groups, 100t, 111t, 118–119
Dyadic interviews, 124–127

Elaboration probes, 47
Elite interviews, 43
Email interviews, 41–42
Emic/etic perspective, 62
Envisioning/planning focus groups, 100t, 111t, 120–122
Epistemological worldview, 2
Epoche, 55
Essence meaning, 55
Essence questions, 64
Essence statements, 57
Ethics, 31–33, 33–34t, 155
Ethnographic research design
  overview, 5–6, 8t
  conversational/discourse analysis and, 89
  data collection and, 20t
  field notes and, 139–140

interview protocols and, 54t, 61–67
process and, 169t
research questions and, 18t, 19t
Exemplars. *See* Case study example
Expert interviews, 43
External audits/inquiry audits, 30

Fetterman, D., 62, 135
Field notes, xxv, 139–140. *See also* Observation rubrics
Finlay, L., 153
Flick, U., 99
Focus group moderator guides
  overview, 12t
  applications, 98
  brainstorming focus groups, 100t, 111t, 120–122
  case study example, 181–183
  considerations for, 101–110
  data transformation for analysis, 128–130
  double-layered designs focus group, 100t
  dual moderators focus groups, 100t, 110t, 116–117
  dueling moderators focus groups, 100t, 111t, 118–119
  envisioning/planning focus groups, 100t, 111t, 120–122
  importance of, 97, 105
  multiple purpose focus groups, 100t
  online/virtual/teleconference focus group, 100t
  piloting of, 127
  pre-focus group profile questionnaires and, 105, 108–109
  program evaluation focus groups, 100t, 111t, 122–124
  single purpose focus groups, 100t, 181–183
  skills needed for, 98–99
  two-way focus groups, 100t, 110t, 111–115
Focus group note-taking recording sheets, 109–110
Focus groups, 96–100, 100t. *See also* Dyadic interviews; Focus group moderator guides
Free association reflective questionnaires, 158, 161

Frey, J. H., 36
Funneling, 45, 127

Gatekeepers, 26–27
Gee, J. P., 69
Generalizability, 27–28, 28t
Gold, R. L., 133
Grounded theory
  overview, 7, 8t
  conversational/discourse analysis and, 89
  data collection and, 21t, 170t
  guiding questions and, 18t
  interview protocols and, 54t
  research questions and, 19t
  theory and, 4, 85
Guba, E. G., 2, 23, 27, 29, 30, 166
Guiding questions, 17, 17–18t

Harris, J., 41
Hermeneutical phenomenology, 56
Historical research design
  overview, 7, 9, 9t
  data collection and, 21t, 170t
  guiding questions and, 18t
  research questions and, 19t
Hodder, I., 143
Horrocks, C., 36
Husserl, Edmund, 55
HyperRESEARCH, 164

Icebreaker questions, 102
Implicit bias, 39–40
Inductive process, 4
Inquiry audits/external audits, 30
Insider-outsider stance. See Positioning; Reflexivity
Institutional/public records, 144t, 146–147. See also Documents and artifacts
Institutional review boards, 33, 33–34t
International Institute for Qualitative Methodology, 191
Interpretive/descriptive research design, 5, 7t, 17t, 19t, 20t, 169t
Interview protocols
  overview, 12t
  considerations for, 44–47
  data transformation for analysis, 82–85
  defined, 37

ethnographic research design and, 54t, 61–67
focus groups compared, 97
life history type, 70, 77–80
narrative dramatism type, 69–70, 75–77
narrative research design and, 54t, 67–80
note-taking recording sheets for, 52–53
phenomenological designs and, 54t, 55–60
piloting of, 80–81
semistructured interviews, 42–44, 50–52, 54t, 179–180
thematic/structural/dialogic type, 69, 72–74
unstructured interviews, 42–44, 47–49, 54t, 177–178
See also Dyadic interviews
Interviews
  applications, 37–38
  defined, 36–37
  focus groups compared, 99
  formats for, 40–42
  skills needed for, 38–40
  types of, 42–44
  See also Dyadic interviews
Introductory questions, 102–103
In vivo coding, 83

Jefferson, G., 87
Joint interviews. See Dyadic interviews
Journals, xxv, 157, 157t, 162–164
Judgments, 25

Key concepts framework, 130
Key informants, 47, 63. See also Ethnographic research design; Interview protocols
Key questions, 103
Kinesics, 52–53
King, N., 36
Krueger, R. A., 96, 128–129
Kutsche, P., 141
Kvale, S., 36

Labov, W., 69
Life histories, 70, 77–80. See also Narrative research design

Lincoln, Y. S., 1, 23, 27, 29, 30, 147, 166
Lived experience, 56–57
Logs/logbooks, xxv, 12t, 87–94. *See also* Notebooks
Long interviews. *See* Unstructured interviews
Lypka, A. E., 88

Marshall, C., 132
Material culture, 143. *See also* Documents and artifacts
Maxwell, J. A., 15
Mayan, M., 31
Member-checking, 29, 142
Memoing, 83
Merriam, S. B., 134, 137
Methodological worldview, 3
Methods triangulation, 29
Moderator's guide, xxv. *See also* Focus group moderator guides
Morgan, D. L., 96, 97, 124
Morse, J. M., 31
Moustakas, C., 56, 57
Multiple purpose focus groups, 100t

Narrative dramatism, 69–70, 75–77
Narrative interviews. *See* Unstructured interviews
Narrative research design
 overview, 6, 8t
 data collection and, 21t
 defining features of, 70–71
 interview protocols, 54t, 67–80
 life history type, 70, 77–80
 narrative dramatism type, 69–70, 75–77
 process and, 170t
 research questions and, 18t, 19t
 thematic/structural/dialogic type, 69, 72–74
 types and approaches, 69–70
Natural setting, 3
Negative case analysis, 30
Nonparticipant observation, 63–64, 135–136
Nonverbals, 52–53
Notebooks, xxv, 162–163. *See also* Logs/logbooks

Note-taking recording sheets, 52–53, 109–110
NVivo, 164

Objectivity, 28t
Observation, 132–136
Observation rubrics, 136–142, 184–185
Oishi, S. M., 36
O'Leary, Z., 146
Olson, K., 31
1-to-1 interviews, 40, 97, 127
Online/virtual/teleconference focus groups, 100t
Open-ended questions, 42, 46–47
Opening questions, 102
Opportunistic strategy, 26

Paired interviews. *See* Dyadic interviews
Participant observation, 63–64, 135–136
Patton, M. Q., 31, 43, 47, 146, 147, 155
Peer debriefing, 29
Peer interviews. *See* Dyadic interviews
Persistent observation, 29
Personal records, 144t, 146–147. *See also* Documents and artifacts
Phenomenological research design
 overview, 5, 8t
 data collection and, 20t
 interview protocols and, 54t, 55–60
 process and, 169t
 research questions and, 17t, 19t
Phenomenology of practice, 57
Photo elicitation, 41
Photovoice, 41
Physical/cultural objects, 144t, 146–147, 149. *See also* Documents and artifacts
Piloting
 conversational/discourse analysis and, 93
 document and artifact rubrics, 151
 focus group moderator guides and, 127
 importance of, 173–174
 interview protocols and, 80–81
 observation rubrics, 141
 observation rubrics and, 141

reflective questionnaires and, 163–164
steps for, 21
Planning/envisioning focus groups, 100t
Positioning
  conversational/discourse analysis and, 88, 89
  ethics and, 32
  observation and, 133, 135–136
Poth, C. N., 23, 154, 155, 166
Practitioner's perspective on research, 167–168
Pre-focus group profile questionnaires, 105, 108–109
Pretests. *See* Piloting
Probes, 47, 104
Professional disciplines, 167
Program evaluation focus groups, 100t, 111t, 122–124
Program evaluation rubrics, 149–150
Prolonged engagement, 29
Protocols, xxvi. *See also* Interview protocols
Proxemics, 52–53
Proximal similarity, 30
Public/institutional records, 144t, 146–147. *See also* Documents and artifacts
Purposeful selection, 26
Purpose statements, 16, 17, 17–18t, 168. *See also* Research questions

Qualitative data. *See* Data sources; *specific qualitative data collection tools*
Qualitative data collection tools
  overview, 11–12, 12–13t
  action plan for, 170–172, 171t, 189–190
  conversation/discourse logs, xxv, 12t, 89–94
  diaries, xxv, 157, 157t, 162–163, 164
  document and artifact rubrics, 12t, 147–151, 186–187
  field notes, xxv, 139–140
  importance of, 168
  journals, 157, 157t
  note-taking recording sheets, 52–53, 109–110
  observation rubrics, 12t, 136–142, 184–185

piloting of, 21
qualitative research designs and, 168–169, 169–170t
recommendations for, 172–174
reflective questionnaires, 156–165, 157t, 188
See also Interview protocols
Qualitative instruments. *See* Qualitative data collection tools
Qualitative interviewing. *See* Interviews
Qualitative interview protocols. *See* Interview protocols
Qualitative observation rubrics, 12t, 136–142, 184–185
The Qualitative Report, 191
Qualitative research
  characteristics of, 3–5
  defined, 1–2
  ethics and, 31–33, 33–34t
  as multifaceted, 166–167
  practitioner's perspective on, 167–168
  as worldview, 2–3
Qualitative Research Consultants Association, 192
Qualitative research designs
  case studies, 6–7, 8t, 18t, 19t, 21t, 170t
  data collection and, 20, 20–21t
  descriptive/interpretive, 5, 7t, 17t, 19t, 20t, 169t
  as emergent, 4
  process of, 168–170, 169–170t
  reflexivity and, 157t
  research questions and, 17–18, 17–19t, 168
  trustworthiness and, 27–31, 28t, 40
  types of, 5–9, 7–9t
  when to use, 9–10
  See also Ethnographic research design; Grounded theory; Historical research design; Narrative research design; Phenomenological research design
Qualitative worldview, 2–3
Questionnaires, 105, 108–109, 156–165, 157t, 188

Index   209

Questions
  design of, 46–47
  focus group moderator guides and, 101–104
  guiding questions, 17, 17–18t
  open-ended questions, 42, 46–47
  reflective questionnaires and, 159
  research questions and, 15–18, 17–19t, 168, 175–176
  sequencing of, 45–46
  types of, 45, 64

Rapport and trust, 52–53
Raw data management, 82–83
Realist ethnography, 61–62
Recording sheets, xxvi, 52–53, 109–110
Reflective questionnaires, 156–165, 157t, 188
Reflexivity
  applications, 154–155
  defined, 153–154
  documents and artifacts and, 146, 148, 150
  observation and, 135
  participant versus researcher, 153
  skills needed for, 155–156
  trustworthiness and, 31
  types of, 156–157, 157t
Reliability, 27–28, 28t
Researcher-position. *See* Positioning; Reflexivity
Researchers
  access to data by, 26–27
  bias and bracketing, 24–25
  as instruments, 4, 23–24
  relationship with participants and, 32, 63–64
  role of, 23–24
Research questions, 15–18, 17–19t, 168, 175–176
Research sites, 26
Rhetorical worldview, 3
Riessman, C. K., 67, 69
Rigor, 15, 31, 173–174
Rogers, R., 88
Rossman, G. B., 132
Rubin, H. J., 36
Rubin, I. S., 36

Rubrics
  case study example, 184–185, 186–187
  defined, xxvi
  document and artifact rubrics, 12t, 147–151, 186–187
  observation rubrics, 12t, 136–142, 184–185
  piloting, 141, 151
  program evaluation rubrics, 149–150
  See also Templates
Ryan, G. W., 43

Sacks, H., 87
Saldana, J., 1
Sampling, 3, 26
Scenario reflective questionnaires, 158, 162
Schegloff, E. A., 87
Secondary coding, 83
Seidman, I., 70
Self-knowledge stance. *See* Reflexivity
Semistructured interviews, 42–44, 50–52, 54t, 179–180
Sequence probes, 47
Sewell, W. H., Jr., 43
Single purpose focus groups, 100t, 181–183
Single topic reflective questionnaires, 158, 160, 188
Snowball strategy, 26
Social engagement questions, 64
Sources of qualitative data, 3, 10, 10–11t, 26–27
Spiers, J., 31
Spradley, J. P., 42, 47, 63
Steering probes, 47
Storytelling. *See* Narrative research design
Structural questions, 64
Structured interviews, 42
Survey platforms, 158–159
Synchronous online interviews, 41

Teleconference/virtual/online focus groups, 100t
Telephone interviews, 41
Templates
  Artifact Rubric for Objects/Tangible Evidence, 149

Artifact Rubric for Researcher Interpretations, 150
Conversational/Discourse Analysis Log, 91–92
Document And Artifact Rubric, 148
Document And Artifact Rubric for Program Assessment/Evaluation, 150
Ethnographic Interview Protocol, 65–67
Focus Group Moderator Guide for Brainstorming/Envisioning, 120–122
Focus Group Moderator Guide for Dual Moderators, 116–117
Focus Group Moderator Guide for Dueling Moderators, 118–119
Focus Group Moderator Guide for Program Evaluation, 122–124
Focus Group Moderator Guide for Two-way Designs, 111–115
Focus Group Moderator Guide: Single Purpose, 106–107
Focus Group Note-Taking Recording Sheet, 109
Focus Group Presession Participant Profile Questionnaire, 108
Interviewer Note-Taking Recording Sheet, 53
Moderator's Guide for Dyadic Interviews, 125–126
Narrative Dramatism Interview Protocol, 75–77
Narrative Life History/Life Story Interview Protocol, 77–80
Narrative Thematic/Structural/Dialogic Interview Protocol, 72–74
Observation Log for Ethnographic Field Notes, 139–140
Observation Rubric for Conceptual Frameworks, 140–141
Observation Rubric for Formal Or Informal Settings, 138
Phenomenological Lived Experience Interview Protocol, 58–60
Reflective Journal And Diary Logs/Notebooks, 162–163

Reflective Questionnaire for Free Association, 161
Reflective Questionnaire for Scenario, 162
Reflective Questionnaire for Single Topic, 160
Semistructured Interview Protocol, 50–51
Unstructured Interview Protocol, 48–49
ten Have, P., 87
Testing alternatives framework, 130
Text message interviews, 41–42
Thematic/structural/dialogic type, 69
Theory triangulation, 29
Thick description, 30, 62
Tools, xxvi. *See also* Qualitative data collection tools
Transcendental phenomenology, 56
Transcripts, 82
Transferability, 28t, 30
Transition questions, 103
Triangulation, 29
Trust and rapport, 52–53
Trustworthiness, 27–31, 28t, 40
Two-person interviews. *See* Dyadic interviews
Two-way focus groups, 100t, 110t, 111–115

Unstructured interviews, 42–44, 47–49, 54t, 177–178

Validity, 27–28, 28t
Van Maanen, J., 56–57, 61, 132
Video recordings, 136
Virtual interviews, 41
Virtual/online/teleconference focus groups, 100t

Walking interviews, 41
Winnowing, 83
Wolcott, H. F., 63
Worldview, 2–3
Wutich, A., 43